Oscillator Circuits

Rudolf F. Graf

Newnes
An Imprint of Elsevier

Boston Oxford Johannesburg Melbourne New Delhi Singapore

I0036745

Newnes is An Imprint of Elsevier

Copyright © 1997 by Butterworth–Heinemann
Copyright © 1992 by Rudolf F. Graf

⊖ A member of the Reed Elsevier group

Permissions may be sought directly from Elsevier's Health Sciences Rights
Department in Philadelphia, USA: phone: (+1)215-238-7869, fax: (+1)215-238-2239,
email: healthpermissions@elsevier.com. You may also complete your request on-line via
the Elsevier Science homepage (http://www.elsevier.com), by selecting 'Customer Support'
and then 'Obtaining Permissions'.

∞ Recognizing the importance of preserving what has been written,
Butterworth–Heinemann prints its books on acid-free paper whenever possible.

Library of Congress Cataloging-in-Publication Data
Graf, Rudolf F.
    [Modern oscillator circuit encyclopedia]
    Oscillator circuits / Rudolf F. Graf
      p.   cm.
    Originally published: The modern oscillator circuit encyclopedia.
  Blue Ridge Summit, PA : TAB Books, c1992.
    Includes index.
    ISBN 0-7506-9883-7 (alk. paper)
    1. Oscillators, Electric. 2. Electronic circuits.
  TK7872.07G68   1996
  621.381'533—dc20   96-43977
                                        CIP

British Library Cataloguing-in-Publication Data
A catalogue record for this book is available from the British Library.

The publisher offers special discounts on bulk orders of this book.
For information, please contact:
Manager of Special Sales
Butterworth–Heinemann
313 Washington Street
Newton, MA 02158–1626
Tel: 617-928-2500
Fax: 617-928-2620

For information on all Newnes electronics publications available, contact
our World Wide Web home page at: http://www.bh.com/bh

Transferred to digital printing 2006

10 9 8 7 6 5 4 3

Contents

Other Books in this Series

Introduction

Like the other volumes in this series, this book contains a wealth of ready-to-use circuits that serve the needs of the engineer, technician, student and, of course, the browser. These unique books contain more practical, ready-to-use circuits focused on a specific field of interest, than can be found anywhere in a single volume.

1

Audio Oscillators

The sources of the following circuits are contained in the Sources section, which begins on page 173. The figure number contained in the box of each circuit correlates to the source entry in the Sources section.

WIEN-BRIDGE SINE-WAVE OSCILLATOR

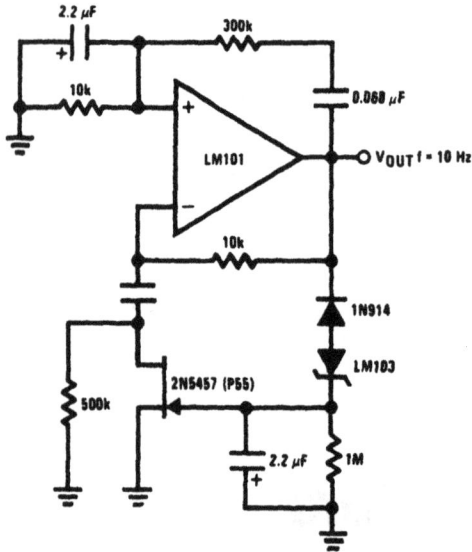

The 2N5457 JFET, as a voltage variable resistor in the amplifier feedback loop, produces a low distortion, constant amplitude sine wave getting the amplifier loop gain just right. The LM103 zener diode provides the voltage reference for the peak sine wave amplitude.

Peak output voltage
$V_p \cong V_z + 1V$

Fig. 1-1

PHASE-SHIFT OSCILLATOR

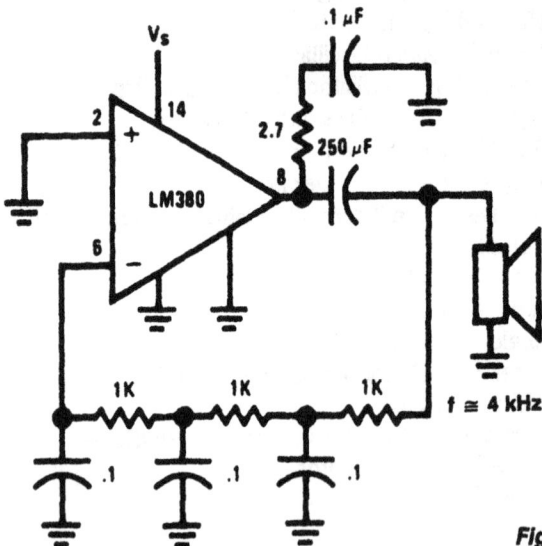

Circuit uses a simple RC network to produce an exceptionally shrill tone from a miniature speaker. With the parts values shown, the circuit oscillates at a frequency of 3.6 kHz and drives a miniature $2^1/_2''$ speaker at ear-piercing volume. The output waveform is a square wave with a width of 150 μs, sloping rise and fall times, and a peak-to-peak amplitude of 4.2 volts (when powered by 9 volts). Current drain of the oscillator is 90 mA at 9 volts, and total power dissipation at this voltage is 0.81 watt, which is well below the 1.25 watts the 14-pin version will absorb (at room temperature) before shutting down.

$f \cong 4$ kHz

Fig. 1-2

WIEN-BRIDGE OSCILLATOR

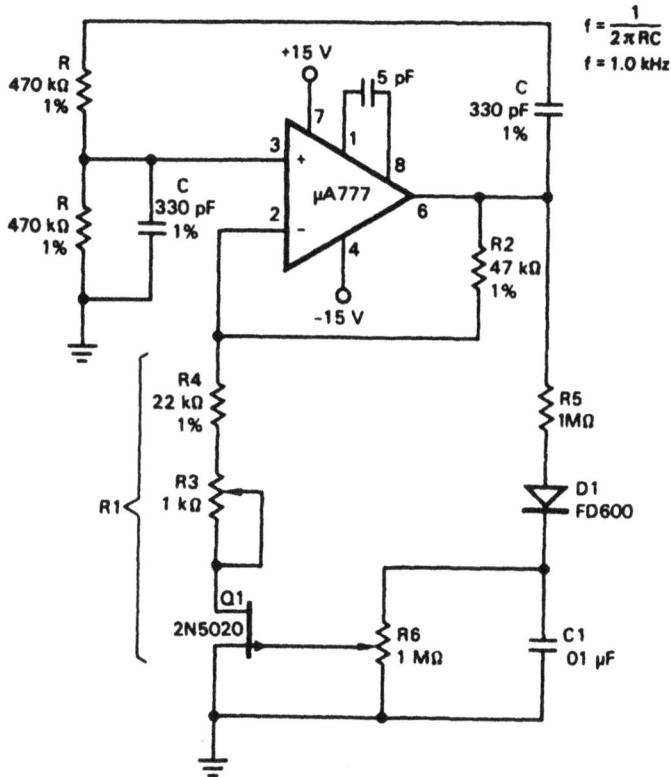

$$f = \frac{1}{2\pi RC}$$

$$f = 1.0 \text{ kHz}$$

FAIRCHILD CAMERA

Fig. 1-3

Field effect transistor, Q1, operates in the linear resistive region to provide automatic gain control. Because the attenuation of the RC network is one-third at the zero phase-shift oscillation frequency, the amplifier gain determined by resistor R2 and equivalent resistor R1 must be just equal to three to make up the unity-gain positive-feedback requirement needed for stable oscillation. Resistors R3 and R4 are set to approximately 1000 ohms less than the required R1 resistance. The FET dynamically provides the trimming resistance needed to make R1 one-half of the resistance of R2. The circuit composed of R5, D1, and C1 isolates, rectifies, and filters the output sine wave, converting it into a dc potential to control the gate of the FET. For the low drain-to-source voltages used, the FET provides a symmetrical linear resistance for a given gate-to-source voltage.

3

CODE-PRACTICE OSCILLATOR

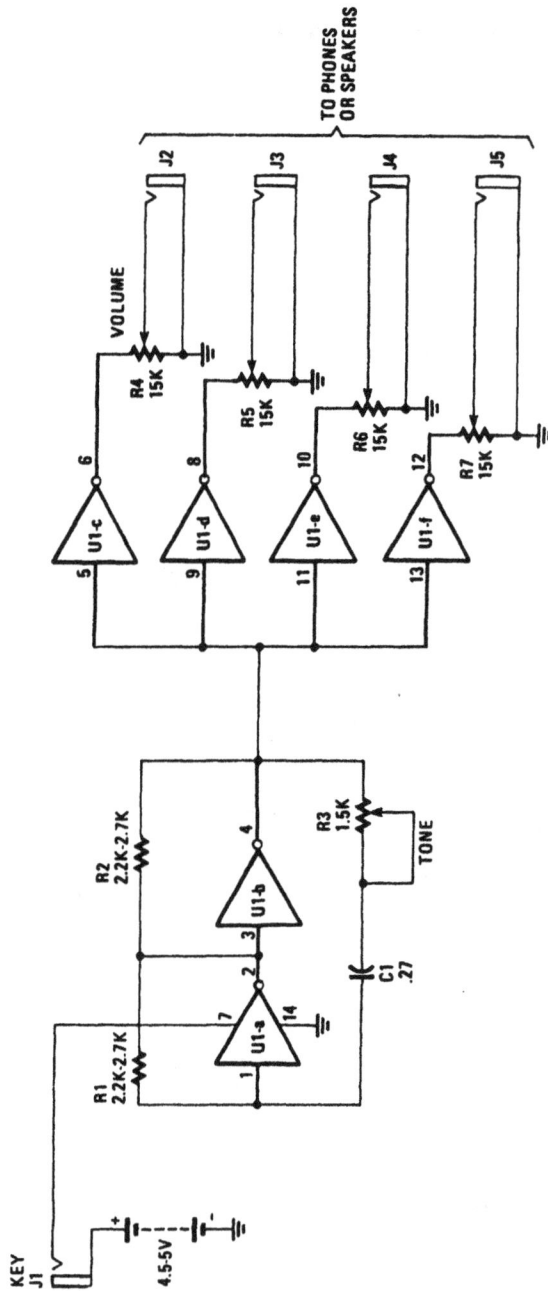

TO PHONES
OR SPEAKERS

Fig. 1-4

The inexpensive 7404 hex-inverter has enough amplification to handle a wide range of transducers. Closing the key completes the battery circuit and applies four to five volts to the 7404. Bias for the first two inverter amps (U1a and U1b) comes from the two resistors, R1 and R2, connected between their inputs and outputs. The capacitor and rheostat (R3/C1) close the feedback loop from the input to the properly phased output. The signal leaving U1b drives the remaining four inverter amplifiers, U1c through U1f; they, in turn, drive the phones or speakers. The volume control potentiometers, R4 – R7, can have any value from 1500 to 10,000 ohms. The smaller values work best when speakers or low impedance phones, are used.

TONE ENCODER

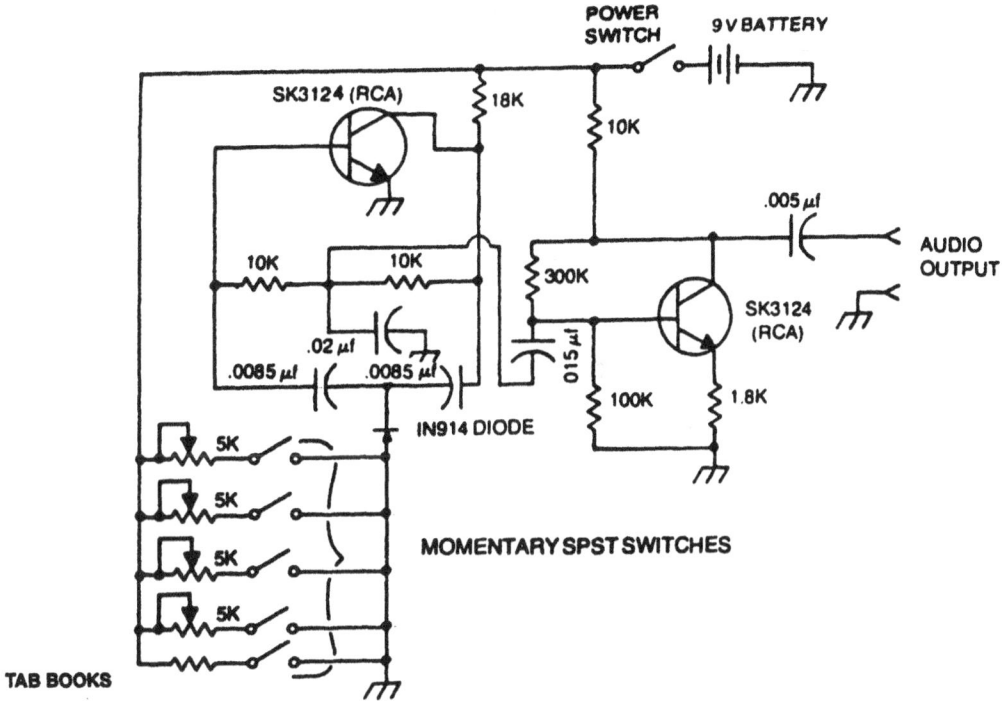

A basic twin-T circuit uses resistors for accurately setting the frequency of the output tones, selected by a pushbutton. Momentary switches produce a tone only when the button is depressed.

FEEDBACK OSCILLATOR

The circuit oscillates because the transistor shifts the phase of the signal 180° from the base to the collector. Each of the RC networks in the circuit is designed to shift the phase 60° at the frequency of oscillation for a total of 180°. The appropriate values of R and C for each network are found from $f = 1/2\sqrt{3\pi RC}$; that equation allows for the 60° phase shift required by the design.

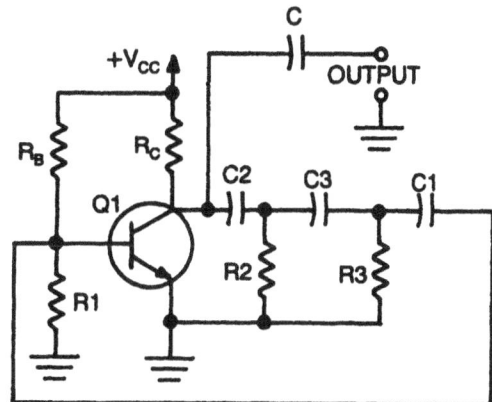

RADIO-ELECTRONICS   Fig. 1-6

WIEN-BRIDGE OSCILLATOR

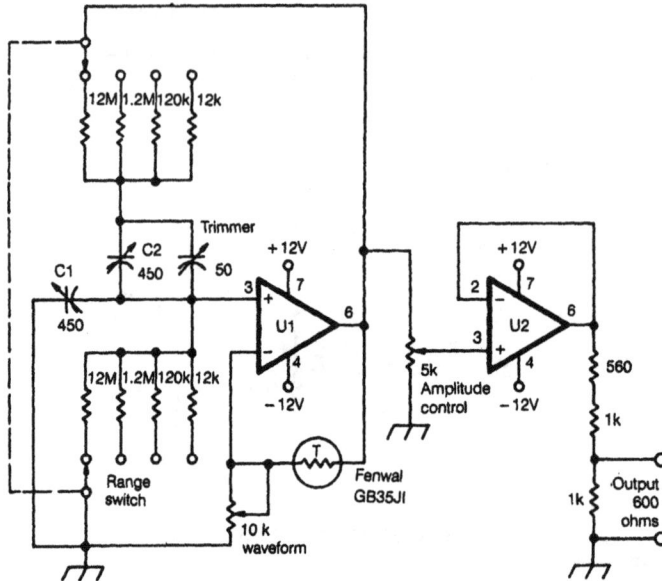

Fig. 1-7

This Wien-bridge sine-wave oscillator, which uses two RCA CA3140 op amps covers 30 Hz to 100 kHz with less than 0.5 percent total harmonic distortion. The 10-kΩ pot is adjusted for the best waveform. Capacitor C1 and C2 is a two-gang, 450-pF variable with its frame isolated from ground. Maximum output into a 600-Ω load is about 1-volt rms.

WIEN-BRIDGE OSCILLATOR

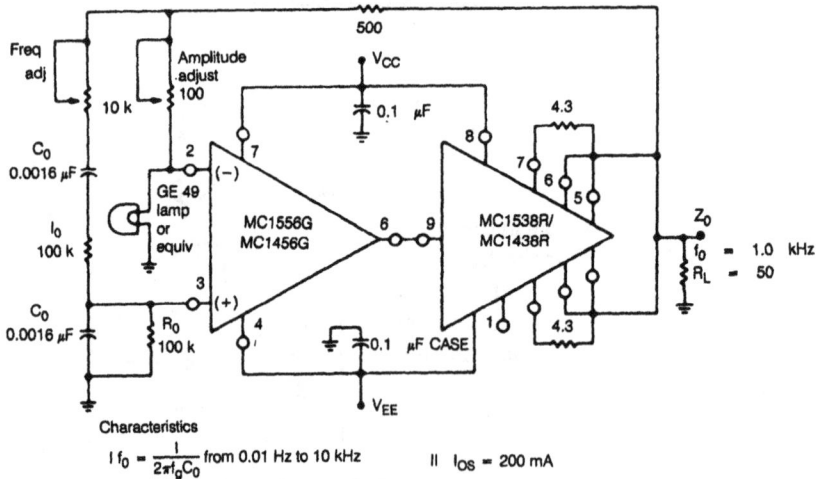

Characteristics

I $f_0 = \frac{1}{2\pi f_0 C_0}$ from 0.01 Hz to 10 kHz   II $I_{OS} = 200$ mA

MOTOROLA

Fig. 1-8

6

PHASE-SHIFT OSCILLATOR

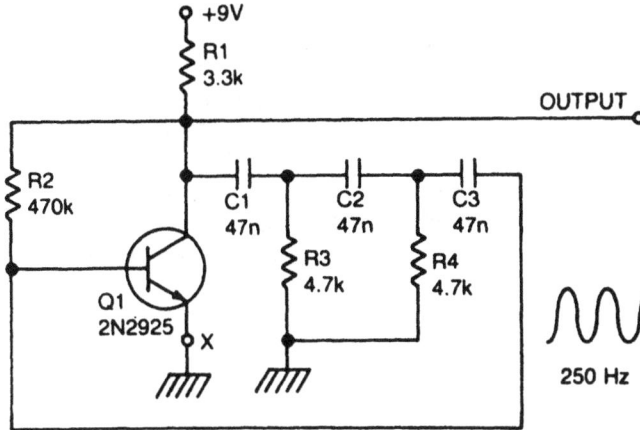

ELECTRONICS TODAY INTERNATIONAL

Fig. 1-9

A single transistor makes a simple phase shift oscillator. The output is a sine wave with distortion of about 10%. The sine wave purity can be increased by putting a variable resistor (25 ohms) in the emitter lead of Q1 (x). The resistor is adjusted so the circuit is only just oscillating, then the sine wave is relatively pure. Operating frequency can be varied by putting a 10 kΩ variable resistor in series with R3, or by changing C1, C2, and C3. Making C1, C2, and C3 equal to 100 nF will halve the operating frequency. Operating frequency can also be voltage controlled by a FET in series with R3, or optically controlled by an LDR in series with R3.

800-Hz OSCILLATOR

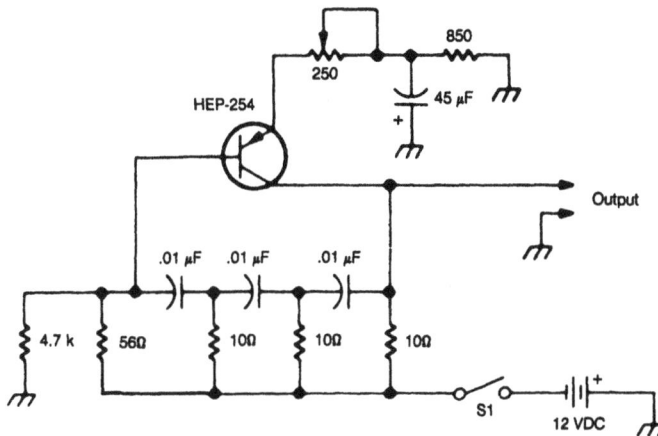

73 AMATEUR RADIO

Fig. 1-10

The following transistors can be used: HEP-254, O.C-2, SK-3004, AT-30H. To increase the frequency, decrease the value of the capacitors in the ladder network.

WIEN-BRIDGE OSCILLATOR

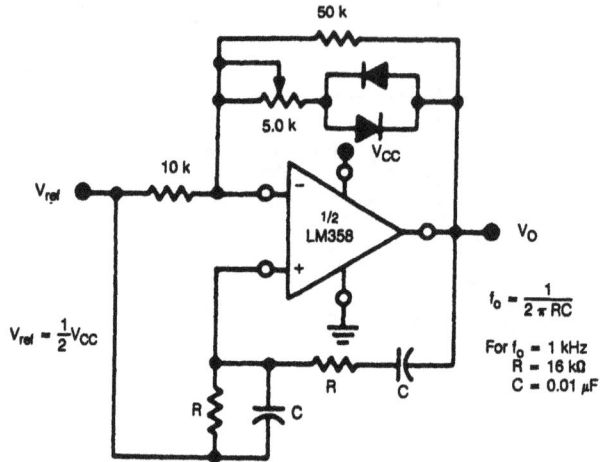

$$f_o = \frac{1}{2 \pi RC}$$

For $f_o = 1$ kHz
R = 16 kΩ
C = 0.01 μF

$V_{ref} = \frac{1}{2} V_{CC}$

Fig. 1-11

MOTOROLA

WIEN-BRIDGE SINE-WAVE OSCILLATOR

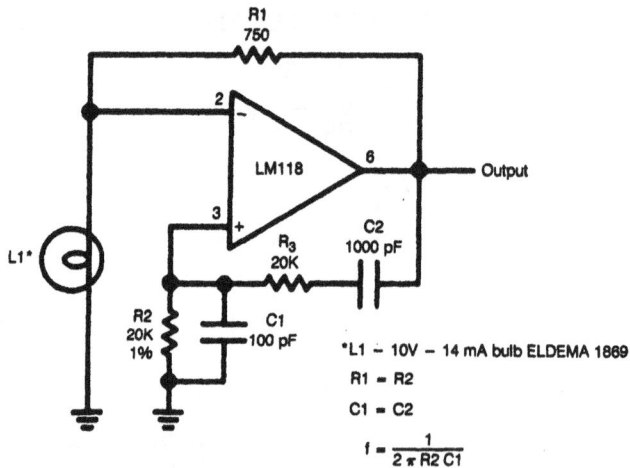

*L1 – 10V – 14 mA bulb ELDEMA 1869

R1 = R2

C1 = C2

$$f = \frac{1}{2 \pi R2\,C1}$$

Fig. 1-12

NATIONAL SEMICONDUCTOR

SINE-WAVE OSCILLATOR

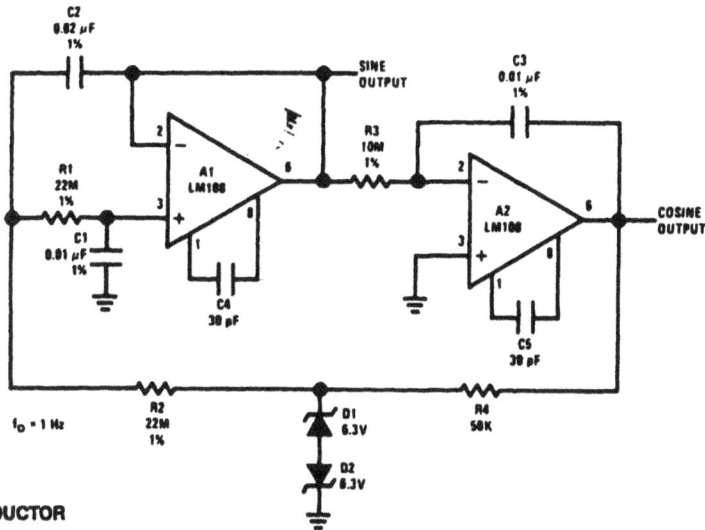

NATIONAL SEMICONDUCTOR

Fig. 1-13

The oscillator delivers a high-purity sinusoid with a stable frequency and amplitude.

EASILY TUNED SINE-/SQUARE-WAVE OSCILLATOR

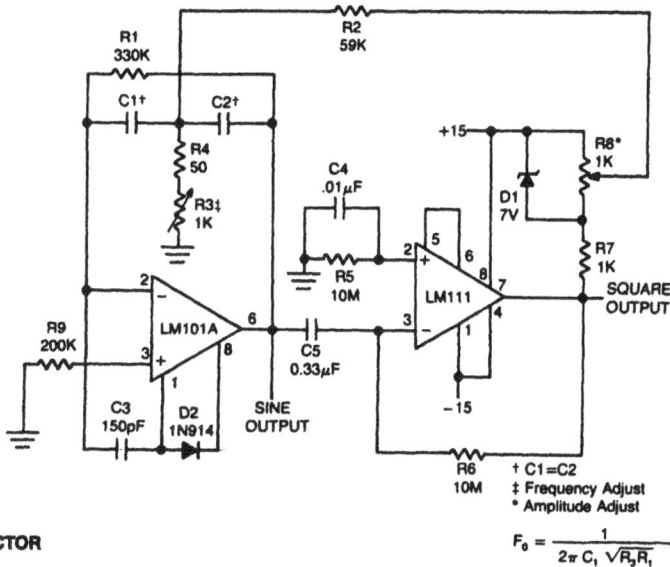

NATIONAL SEMICONDUCTOR

† C1=C2
‡ Frequency Adjust
* Amplitude Adjust

$$F_o = \frac{1}{2\pi C_1 \sqrt{R_2 R_1}}$$

Fig. 1-14

This circuit will provide both a sine- and square-wave output for frequencies from below 20 Hz to above 20 kHz. The frequency of oscillation is easily tuned by varying a single resistor.

VERY-LOW-FREQUENCY GENERATOR

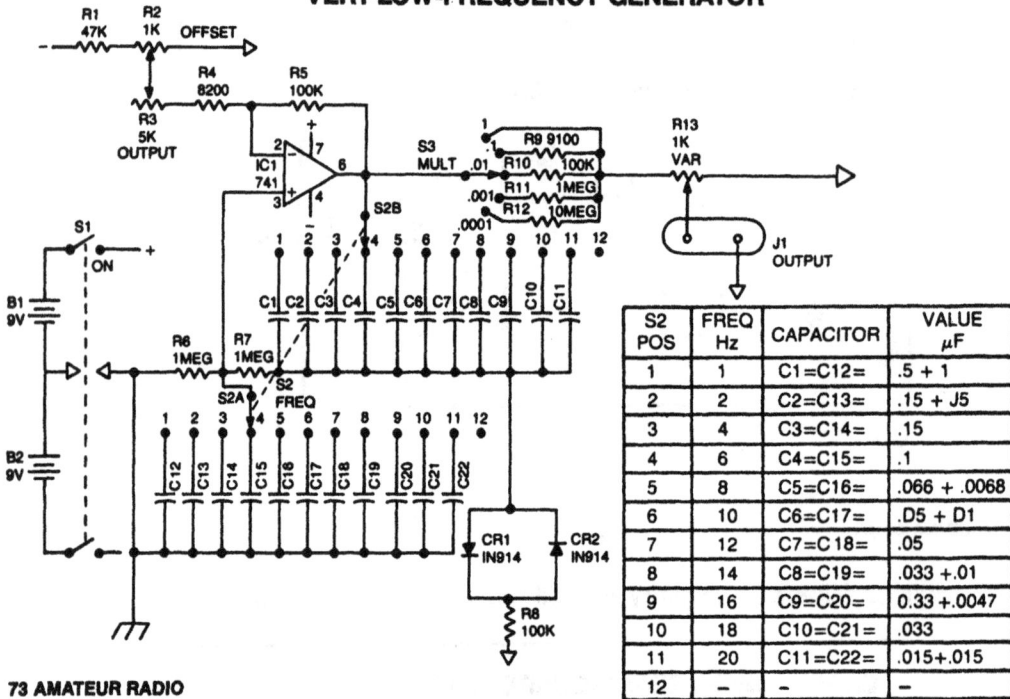

R1 47K R2 1K OFFSET
R4 8200 R5 100K
R3 5K OUTPUT
IC1 741
S3 MULT
R9 9100 .01 R10 100K
.001 R11 1MEG
.0001 R12 10MEG
R13 1K VAR
S2B
S1 ON
B1 9V
R6 1MEG R7 1MEG
S2A S2 FREQ
B2 9V
C1 C2 C3 C4 C5 C6 C7 C8 C9 C10 C11
C12 C13 C14 C15 C16 C17 C18 C19 C20 C21 C22
CR1 IN914 CR2 IN914
R8 100K
J1 OUTPUT

S2 POS	FREQ Hz	CAPACITOR	VALUE μF
1	1	C1=C12=	.5 + 1
2	2	C2=C13=	.15 + J5
3	4	C3=C14=	.15
4	6	C4=C15=	.1
5	8	C5=C16=	.066 + .0068
6	10	C6=C17=	.D5 + D1
7	12	C7=C18=	.05
8	14	C8=C19=	.033 +.01
9	16	C9=C20=	0.33 +.0047
10	18	C10=C21=	.033
11	20	C11=C22=	.015+.015
12	–	–	–

Fig. 1-15

This Wien-bridge oscillator generates frequencies of 1 Hz and 2 to 20 Hz in 2-Hz steps. Maximum output amplitude is 3-volts rms or 8.5 volts peak-to-peak. A pot-and-switch attenuator allows the output level to be set with a fair degree of precision to any value within a range of 5 decades.

AUDIO OSCILLATOR

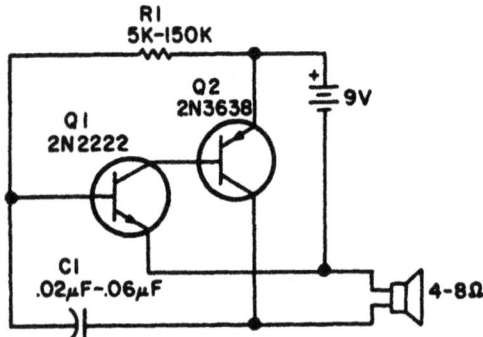

R1 5K-150K
Q2 2N3638
9V
Q1 2N2222
C1 .02μF-.06μF
4-8Ω

Almost any transistor will work. R1 and C1 will vary the tone.

Fig. 1-16

ELECTRONIC BAGPIPE

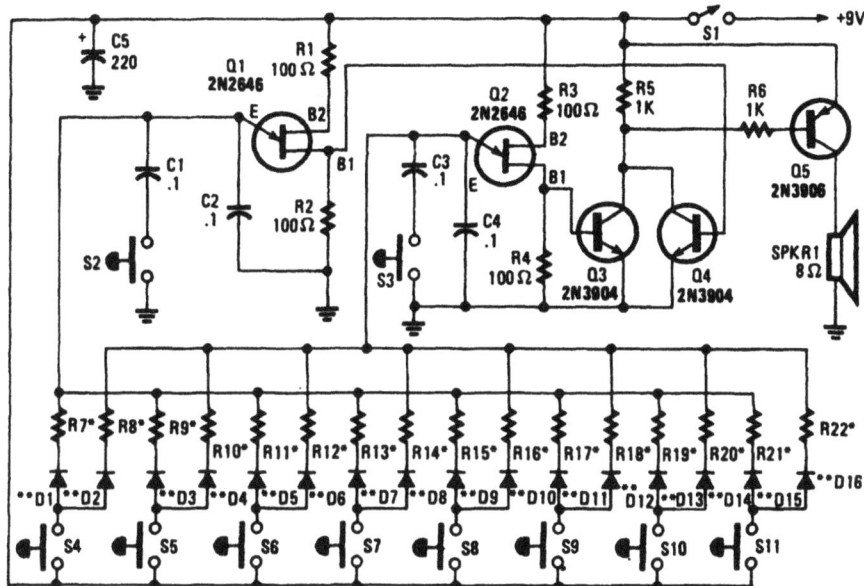

Fig. 1-17

This circuit mimics the dual-tone drone sound that's produced by the unusual wind instrument. Unijunction transistors Q1 and Q2 are connected in similar audio-oscillator circuits. Each of the oscillator frequencies is determined by one of the two resistors selected by one of the pushbutton switches, S4 through S11. The odd-numbered resistors in R7 to R21, determine the frequency for the Q1 oscillator circuit and the even-numbered resistors in R8 through R22, determine the frequency for Q2's circuit.

When S4 is pressed, the positive supply is connected to both R7 and R8 through isolation diodes D1 and D2, causing both oscillators to operate. A narrow, fast-rising positive pulse is developed at B1 of both Q1 and Q2 for each cycle of operation. Transistors Q3 and Q4 serve as a simple audio mixer, which is used to combine the pulses from each oscillator. The mixed signal at the collectors of Q3 and Q4 is coupled through R6 to the base of Q5, which amplifies and drives an 8-Ω speaker, SPKR1. Switches, S2 and S3 are used to reduce the oscillator's frequency by about 50% when closed, to produce a new group of tones.

SIMPLE AUDIO SINE-WAVE GENERATOR

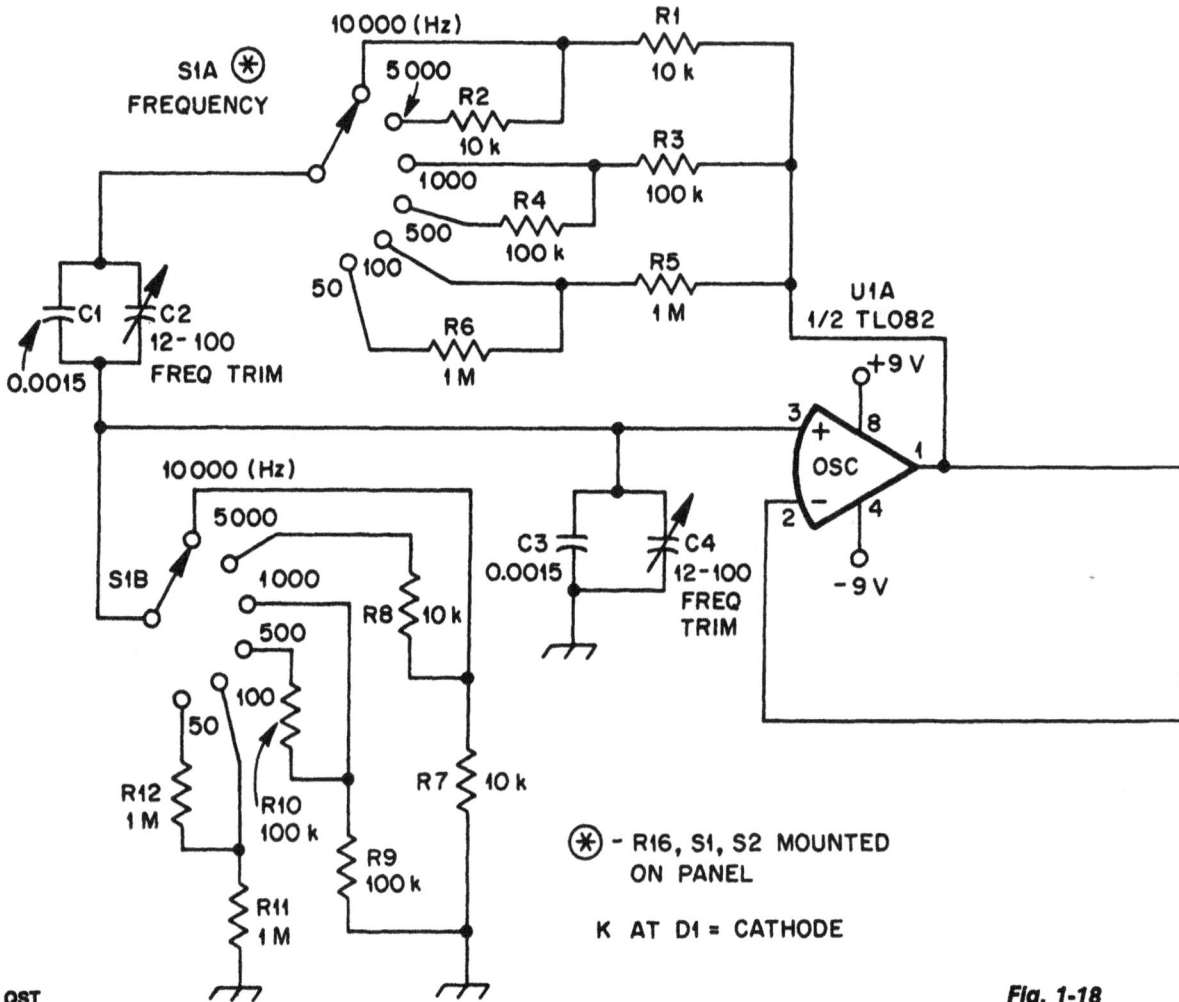

Fig. 1-18

U1A, an op amp, oscillates at the frequency at which the phase shift in the Wien-bridge network is exactly zero degrees. Changing bridge component values changes the oscillator frequency. In this circuit, we need change only the two resistors to do this. S1A chooses a value among R1 through R6, and S1B similarly selects a value from R7 through R12. U1A must provide enough gain to overcome losses in the bridge, but not so much gain that oscillation builds up to the point of overload and distortion. U2 and C1 automatically regulate circuit gain to maintain oscillation. U2 places D1 across R13 with the proper polarity

S2A ✳

+9 V
TO U1-8

R19 10

ON   OFF

BT1
9 V

C5 0.1   C6 100 µF 16 V

−9 V
TO U1-4

R20 10

S2B   BT2

C7 0.1   C8 100 µF 16 V

9 V

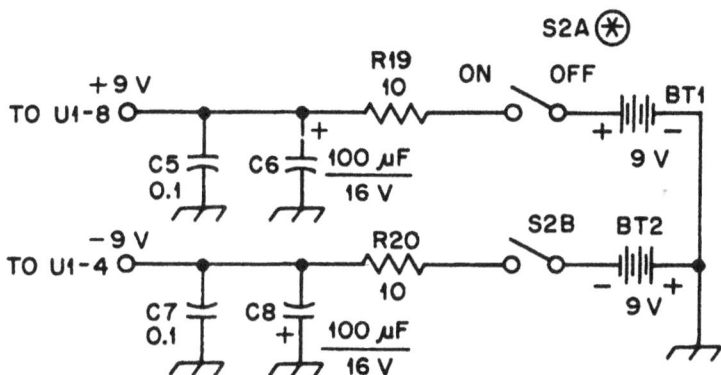

EXCEPT AS INDICATED, DECIMAL
VALUES OF CAPACITANCE ARE
IN MICROFARADS (µF); OTHERS
ARE IN PICOFARADS (pF);
RESISTANCES ARE IN OHMS;
k = 1000, M = 1000 000.

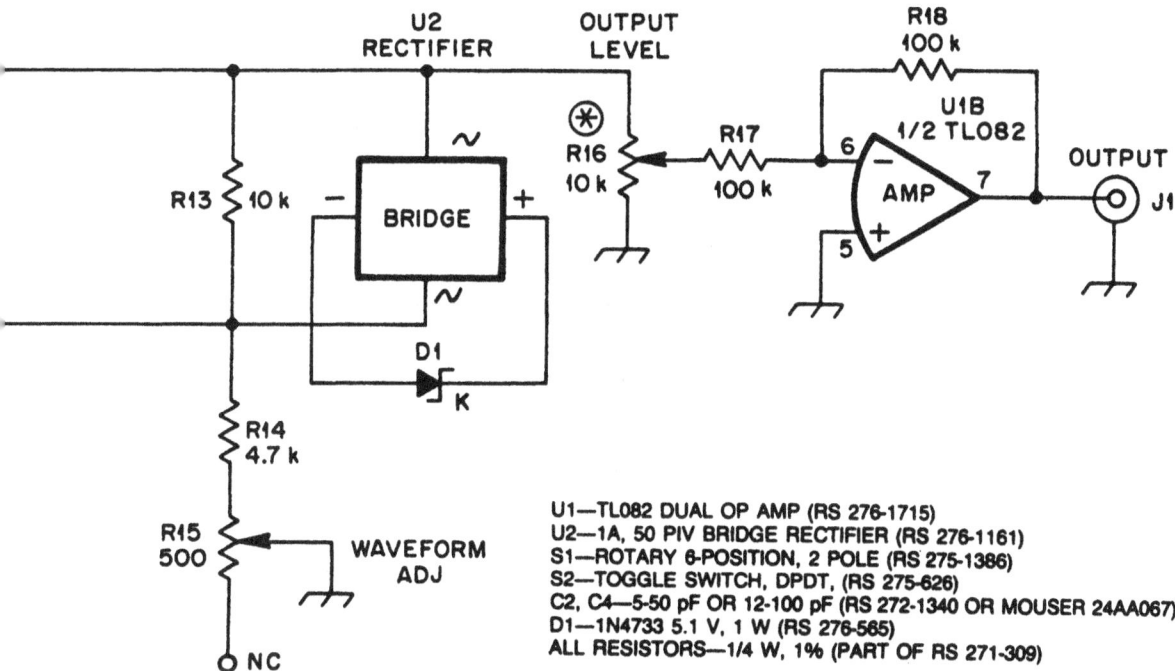

U2 RECTIFIER

OUTPUT LEVEL

R18 100 k

✳ R16 10 k

R17 100 k

6 −

U1B 1/2 TL082

AMP

7

OUTPUT J1

R13 10 k

BRIDGE

5 +

D1
K

R14 4.7 k

R15 500

WAVEFORM ADJ

NC

U1—TL082 DUAL OP AMP (RS 276-1715)
U2—1A, 50 PIV BRIDGE RECTIFIER (RS 276-1161)
S1—ROTARY 6-POSITION, 2 POLE (RS 275-1386)
S2—TOGGLE SWITCH, DPDT, (RS 275-626)
C2, C4—5-50 pF OR 12-100 pF (RS 272-1340 OR MOUSER 24AA067)
D1—1N4733 5.1 V, 1 W (RS 276-565)
ALL RESISTORS—1/4 W, 1% (PART OF RS 271-309)

on both positive and negative alterations of the signal at pin 1 of U1. As the voltage at pin 1 of U1 approaches its peak value, D1 enters its Zener breakdown region, effectively shunting R13 with a resistive load. This shunting increases the amount of negative feedback around U1, reducing its gain. R15, Waveform Adj, allows you to optimize circuit operation for lowest distortion. U1B provides isolation between oscillator and load. With the values shown for R17 and R18, U1B operates at unity gain.

LOW-COST WIEN-BRIDGE OSCILLATOR

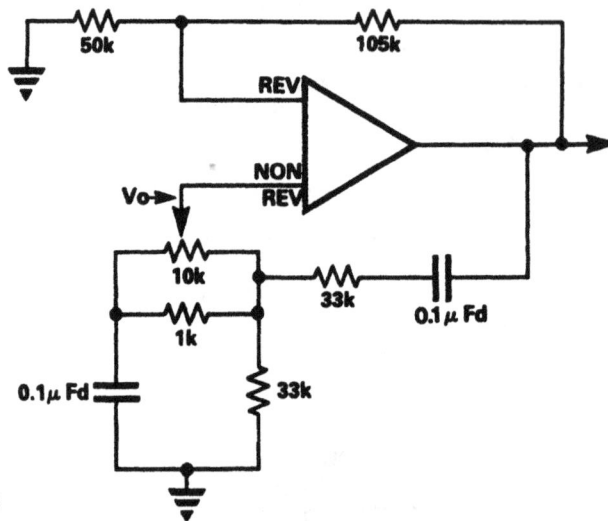

ELECTRONIC ENGINEERING

Fig. 1-19

In the circuit, the frequency-trimming component is arranged so that the voltage across it is in quadrature with the voltage V_o from the bridge; as it is adjusted the attenuation of the bridge only changes a little, avoiding the need for a two-gang component. The range of variation of frequency is very limited. By using a high-gain amplifier and metal-film feedback resistors the loop gain can be set so that the unit just oscillates and the use of an automatic-gain-setting component (a thermistor, for example) is eliminated.

MODIFIED UJT RELAXATION OSCILLATOR

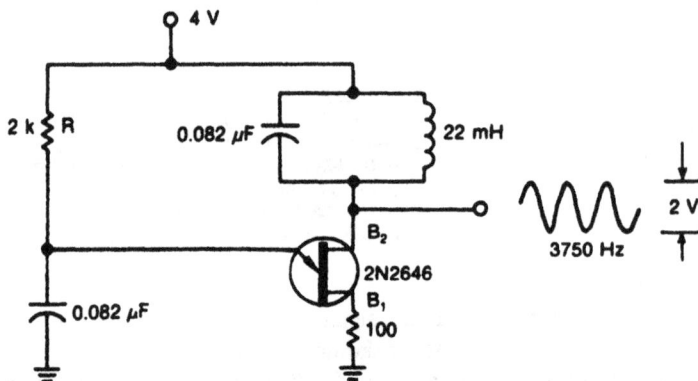

ELECTRONIC DESIGN

Fig. 1-20

By placing a tuned circuit in the UJT oscillator's current-pulse path, a 3750-Hz sinusoid can be created at B2 with the component values shown.

A 555 RC AUDIO OSCILLATOR

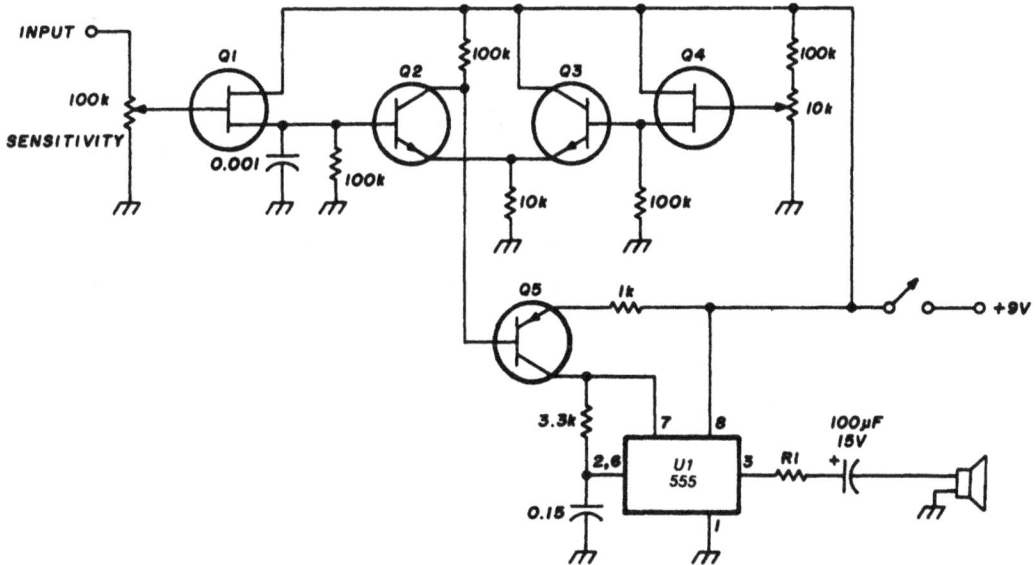

HAM RADIO

Fig. 1-21

Transistor Q5 and the 1000-Ω resistor form the variable element that is needed for controlling the frequency of VCO by limiting the charging current flowing into the 0.15-µF timing capacitor according to the forward bias being applied to Q5. As the voltage on pins 2 and 6 of U1 reach $2/3$ V_{CC} (about 6 volts with a 9-volt supply) the timer will fire and pin 3 will be pulled low. Pin 7, an open collector output, goes low and begins to discharge the timing capacitor—through the 3.3-kΩ resistor. The discharge time provided by this resistor assures a reasonable, although asymmetrical, waveform for the aural signal generated by U1. At $1/3$ V_{CC}, the internal flip-flop resets, the output on pin 3 goes high, the open collector output on pin 7 floats, and the timing cycle begins again.

WIEN-BRIDGE OSCILLATOR USES CMOS CHIP

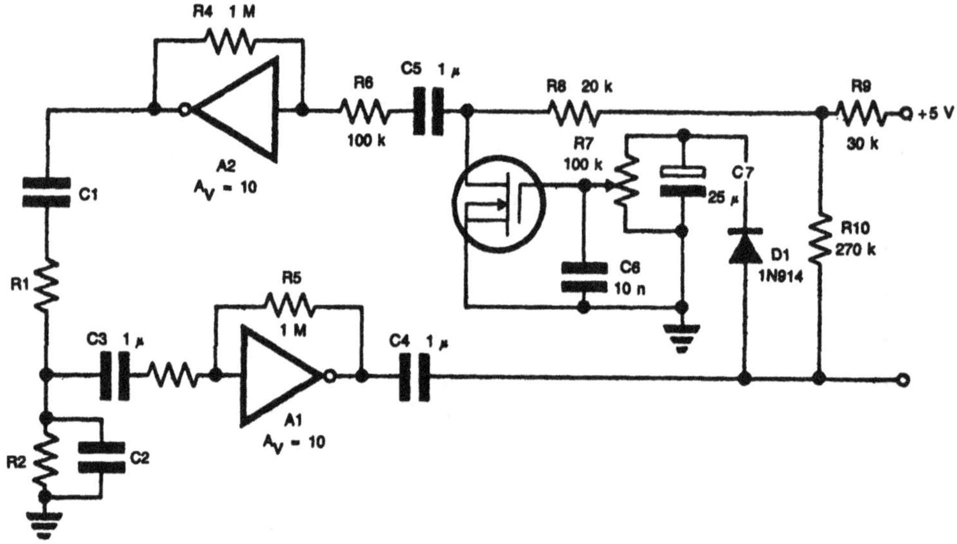

ELECTRONIC ENGINEERING

Fig. 1-22

ADJUSTABLE SINE-WAVE AUDIO OSCILLATOR

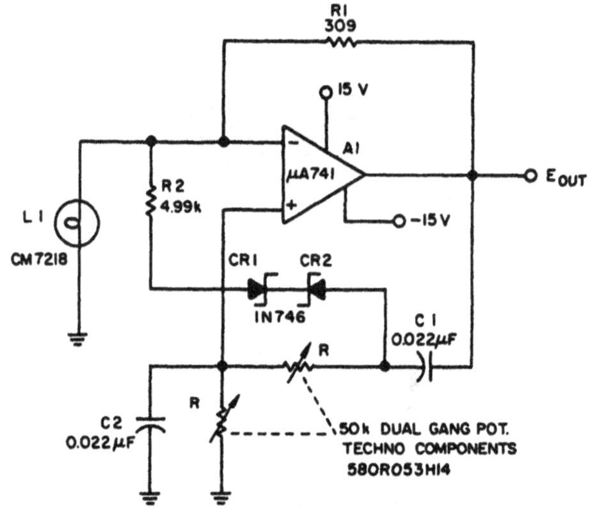

ELECTRONIC DESIGN

Fig. 1-23

Waveform purity at low frequencies for a Wien-bridge oscillator is enhanced by diode limiting. Lamp L1 stabilizes the loop gain at higher frequencies while the limiting action of R2, CR1, and CR2 prevents clipping at low frequencies and increases the frequency adjustment range from about 3:1 to greater than 10:1.

ONE-IC AUDIO GENERATOR

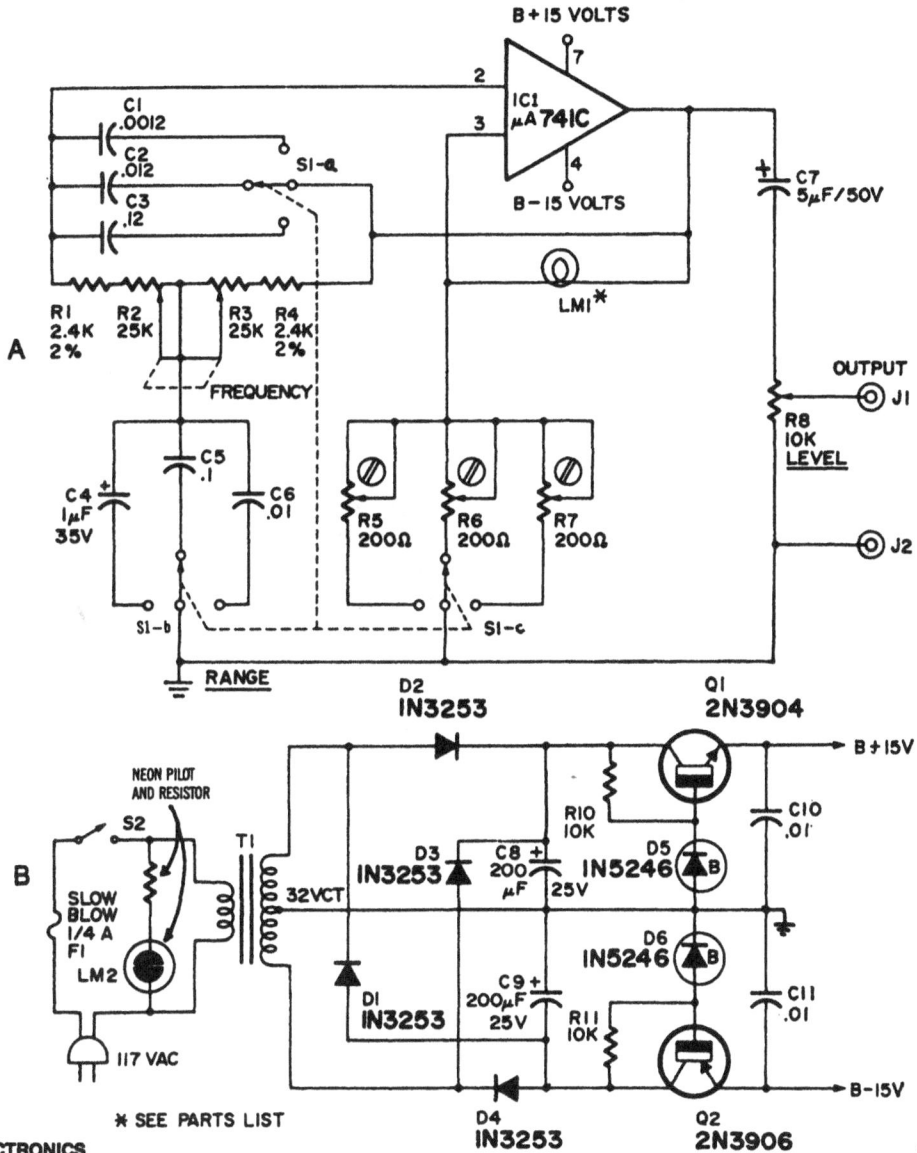

B + 15 VOLTS

2

IC1
µA 741C

3

B – 15 VOLTS

7

4

C1 .0012
C2 .012
C3 .12

S1-a

R1 2.4K 2% R2 25K R3 25K R4 2.4K 2%

A

FREQUENCY

C4 1µF 35V   C5 .1   C6 .01

S1-b

RANGE

R5 200Ω   R6 200Ω   R7 200Ω

S1-c

LM1*

C7 5µF/50V

OUTPUT
J1

R8 10K
LEVEL

J2

D2 1N3253

Q1 2N3904

B + 15V

NEON PILOT AND RESISTOR

S2

R10 10K

C10 .01

D5 1N5246

B

SLOW BLOW 1/4 A F1

LM2

T1

32VCT

D3 1N3253   C8 + 200 µF 25V

D1 1N3253   C9 + 200µF 25V

D6 1N5246

C11 .01

R11 10K

117 VAC

* SEE PARTS LIST

D4 1N3253

D1 1N3253

Q2 2N3906

B – 15V

Fig. 1-24

This high-quality low-cost generator covers 20 Hz to 20 kHz in three bands with less than 1% distortion. LM1—10 V, 14 mA (344, 1869, 914) or 10 V, 10 mA (913, 367).

A = oscillator
B = power supply

SIMPLE TWO-TONE GENERATOR

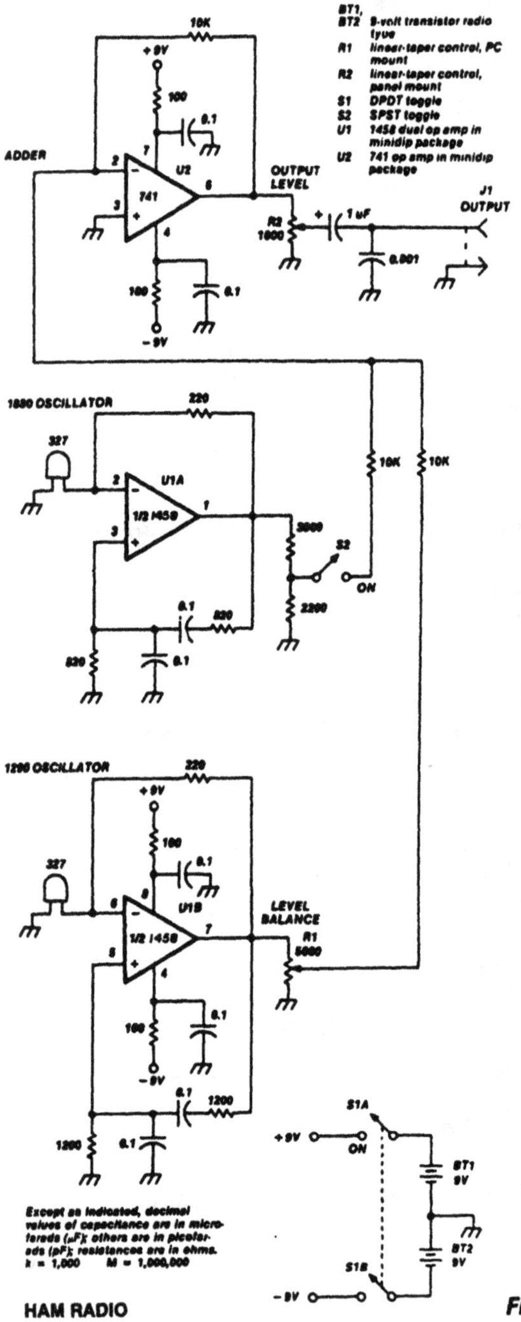

Two 741 operational amplifiers are used for the active elements in this Wien-bridge oscillator (the 1458 is the dual version of the 741). Frequencies of the two oscillators were chosen to fit standard component values. Other frequencies between 500 and 2000 Hz can be employed. They should not be harmonically related. The output level of U1A is set by a resistive divider, while the output of U1B is adjustable through R1. The output of the two oscillators is combined in U2, an op-amp adder with unity gain. The output from U2 can be adjusted using R2.

Fig. 1-25

HAM RADIO

LOW-DISTORTION THERMALLY STABILIZED WIEN-BRIDGE OSCILLATOR

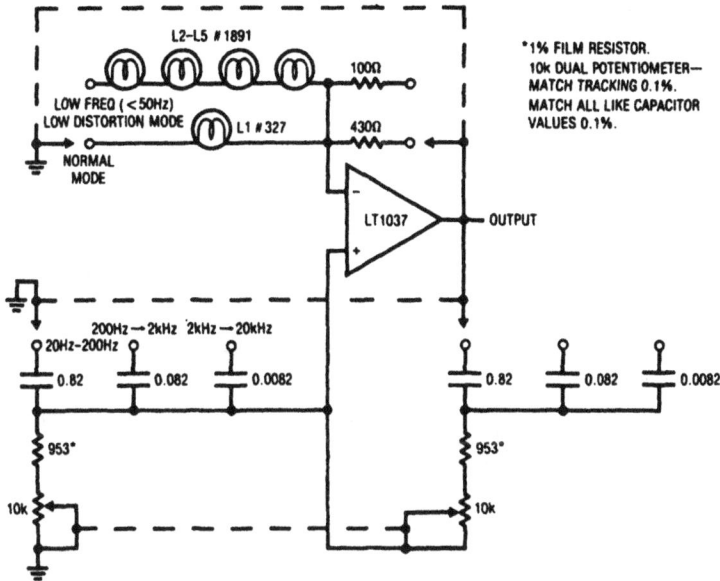

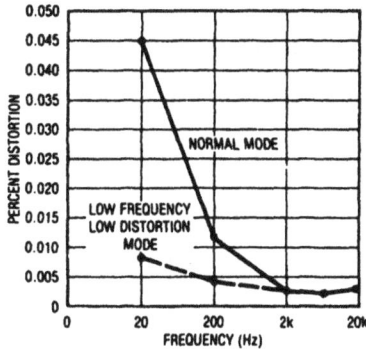

L2–L5 # 1891

100Ω

*1% FILM RESISTOR.
10k DUAL POTENTIOMETER—
MATCH TRACKING 0.1%.
MATCH ALL LIKE CAPACITOR
VALUES 0.1%.

LOW FREQ (<50Hz)
LOW DISTORTION MODE

L1 # 327

430Ω

NORMAL
MODE

LT1037

OUTPUT

200Hz → 2kHz   2kHz → 20kHz
20Hz–200Hz
20Hz–200Hz

0.82   0.082   0.0082

0.82   0.082   0.0082

953*

953*

10k

10k

Oscillator Distortion vs Frequency

PERCENT DISTORTION

0.050
0.045
0.040
0.035
0.030
0.025
0.020
0.015
0.010
0.005
0

NORMAL MODE

LOW FREQUENCY
LOW DISTORTION
MODE

0   20   200   2k   20k
FREQUENCY (Hz)

LINEAR TECHNOLOGY CORP.

Fig. 1-26

A variable Wien bridge provides frequency tuning from 20 Hz to 20 kHz. Gain control comes from the positive temperature coefficient of the lamp. When power is applied, the lamp is at a low resistance value, the gain is high, and oscillation amplitude builds. The lamp's gain-regulating behavior is flat within 0.25 dB over the 20 Hz – 20 kHz range of the circuit. Distortion is below 0.003%. At low frequencies, the thermal time constant of the small normal-mode lamp begins to introduce distortion levels about 0.01%. This is because of *hunting* when the oscillator's frequency approaches the lamp's thermal time constant. This effect can be eliminated, at the expense of reduced output amplitude and longer amplitude settling time, by switching to the low-frequency, low-distortion mode. The four large lamps give a longer thermal time constant, and distortion is reduced.

AUDIO OSCILLATOR

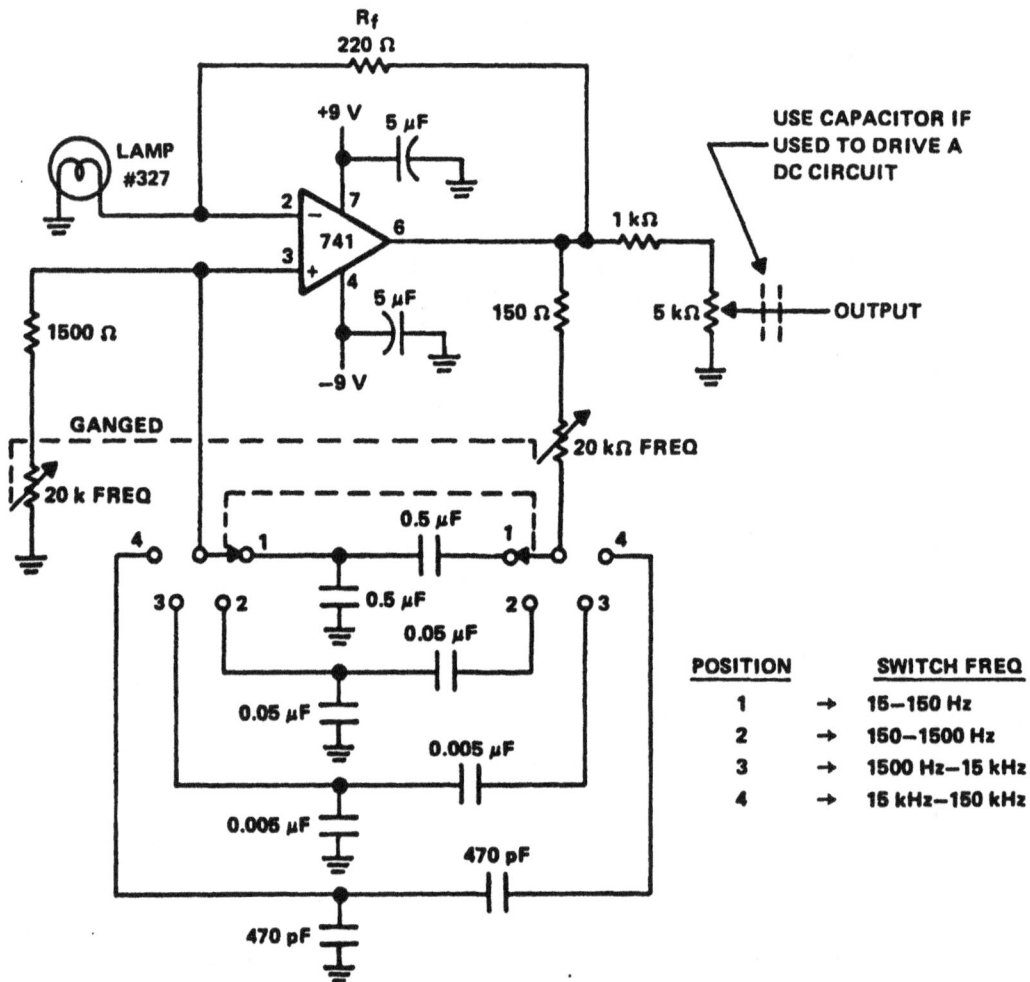

POSITION		SWITCH FREQ
1	→	15–150 Hz
2	→	150–1500 Hz
3	→	1500 Hz–15 kHz
4	→	15 kHz–150 kHz

AUDIO OSCILLATOR (continued)

A Wien-bridge oscillator produces sine waves with very low distortion level. The Wien-bridge oscillator produces zero phase shift at only one frequency ($f = 1/2\pi RC$), which will be the oscillation frequency. Stable oscillation can occur only if the loop gain remains at unity at the oscillation frequency. The circuit achieves this control by using the positive temperature coefficient of a small lamp to regulate gain (R_f/R_{LAMP}) as the oscillator attempts to vary its output. The oscillator shown here has four frequency bands covering about 15 Hz to 150 kHz. The frequency is continuously variable within each frequency range with ganged 20-kΩ potentiometers. The oscillator draws only about 4.0 mA from the 9-V batteries. Its output is from 4 to 5 V with a 10-kΩ load and the R_f (feedback resistor) is set at about 5% below the point of clipping. As shown, the center arm of the 5-kΩ output potentiometer is the output terminal. To couple the oscillator to a dc-type circuit, a capacitor should be inserted in series with the output lead.

AUDIO GENERATOR

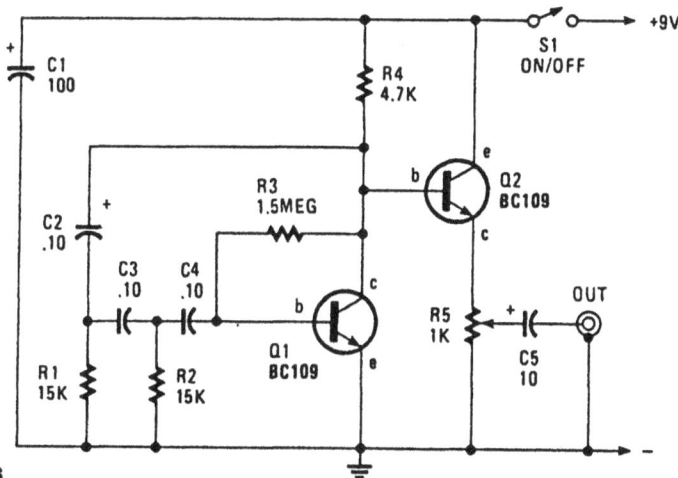

HANDS-ON ELECTRONICS

Fig. 1-28

This circuit produces a sinusoidal output of about 8 V pk-pk, which can be varied down to zero, at about 500 Hz. The signal is generated by a phase-shift oscillator.

SINGLE-SUPPLY WIEN-BRIDGE OSCILLATOR

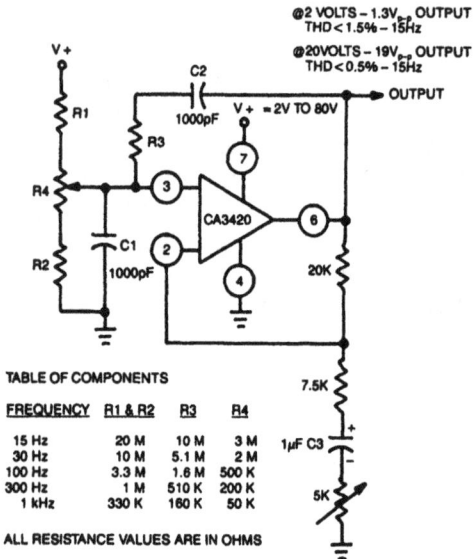

@2 VOLTS – 1.3V$_{p-p}$ OUTPUT
THD < 1.5% – 15Hz

@20VOLTS – 19V$_{p-p}$ OUTPUT
THD < 0.5% – 15Hz

OUTPUT

V + = 2V TO 80V

TABLE OF COMPONENTS

FREQUENCY	R1 & R2	R3	R4
15 Hz	20 M	10 M	3 M
30 Hz	10 M	5.1 M	2 M
100 Hz	3.3 M	1.6 M	500 K
300 Hz	1 M	510 K	200 K
1 kHz	330 K	160 K	50 K

ALL RESISTANCE VALUES ARE IN OHMS

The adjustment of R4 contributes to the comparatively symmetrical output transfer characteristic of the CA3420 BiMOS op amp. To extend the lower operating frequency, remove C3 and use a dual supply.

GE/RCA

Fig. 1-29

SUPER-LOW-DISTORTION VARIABLE SINE-WAVE OSCILLATOR

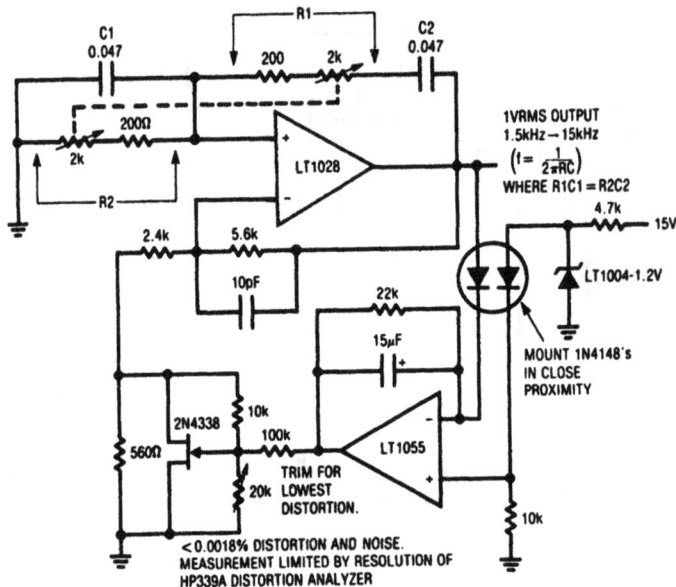

1VRMS OUTPUT
1.5kHz – 15kHz

$\left(f = \frac{1}{2\pi RC} \right)$

WHERE R1C1 = R2C2

MOUNT 1N4148's
IN CLOSE
PROXIMITY

TRIM FOR
LOWEST
DISTORTION.

< 0.0018% DISTORTION AND NOISE.
MEASUREMENT LIMITED BY RESOLUTION OF
HP339A DISTORTION ANALYZER

LINEAR TECHNOLOGY CORP.

Fig. 1-30

1-kHz OSCILLATOR

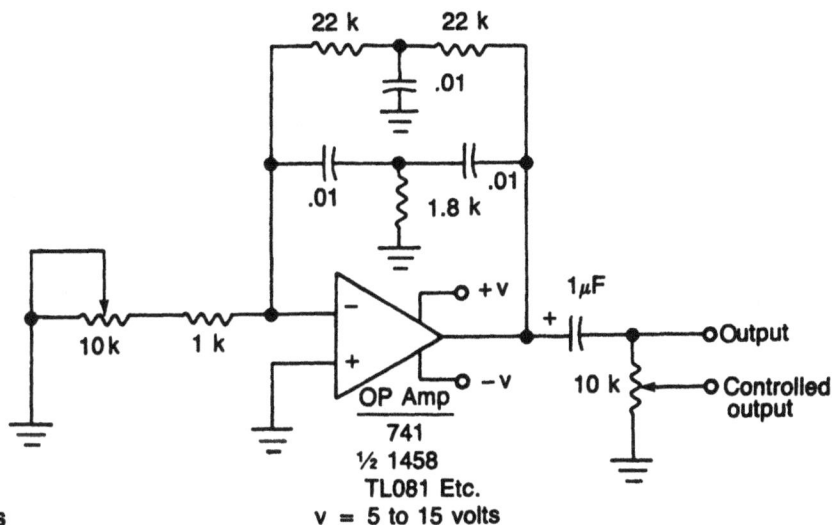

WILLIAM SHEETS

v = 5 to 15 volts

Fig. 1-31

If fine output control is desired, add the 10-kΩ potentiometer. When the oscillator is connected to a dc circuit, then connect a dc-blocking capacitor in series with the potentiometer's wiper arm.

INEXPENSIVE OSCILLATOR IS TEMPERATURE STABLE

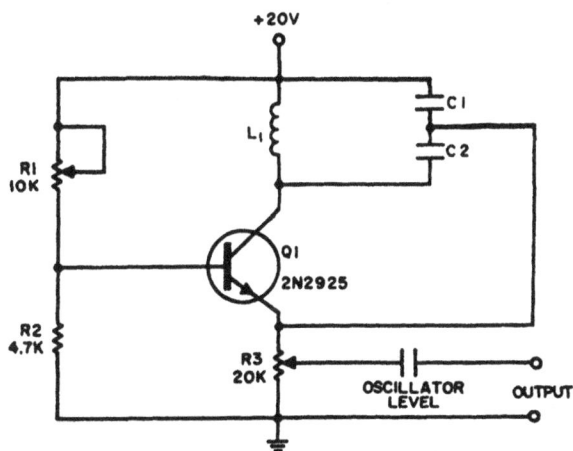

ELECTRONIC DESIGN

Fig. 1-32

The Colpitts sinusoidal oscillator provides stable output amplitude and frequency from 0°F to +150°F. In addition, output amplitude is large and harmonic distortion is low: Oscillation is sustained by feedback from the collector tank circuit to the emitter. The oscillator's frequency is determined by:

$$f = \cfrac{1}{2\pi \sqrt{\cfrac{L_1 C_1 C_2}{C_1 + C_2}}}$$

Potentiometer R3 is an output level control. Control R1 can be used to adjust base bias for maximum-amplitude output. The circuit was operated at 50 kHz with L1=10mH, C1=3500 pF, and C2=1500 pF.

CODE-PRACTICE OSCILLATOR

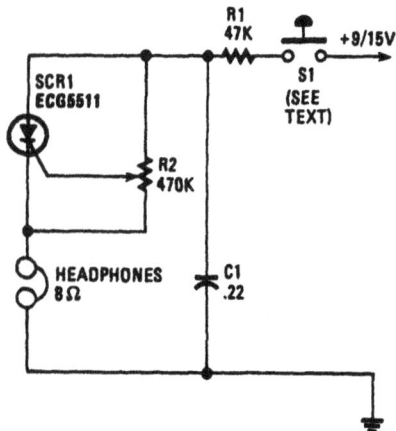

Capacitor C1 charges through resistor R1, and when the gate level established by potentiometer R2 is high enough, the SCR is triggered. Current flows through the SCR and earphones, discharging C1. The anode voltage and current drop to a low level, so the SCR stops conducting and the cycle is repeated. Resistor R2 lets the gate potential across C1 be adjusted, which charges the frequency or tone. Use a pair of 8-Ω headphones. The telegraph key goes right into the B+ line, 9-V battery.

HANDS-ON ELECTRONICS/
POPULAR ELECTRONICS

Fig. 1-33

AUDIO OSCILLATOR

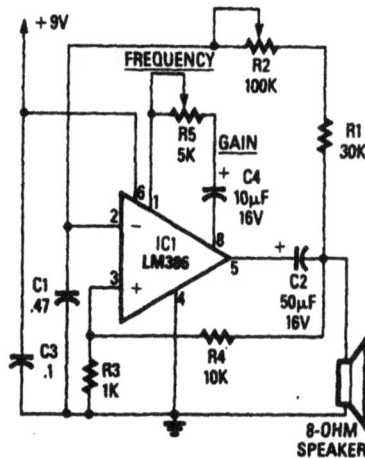

Reprinted with permission of Radio-Electronics Magazine, August 1986. Copyright Gernsback Publications, Inc., 1986.

Fig. 1-34

The circuit's frequency of oscillation is $f = 2.8/[C_1 \times (R_1 + R_2)]$. Using the values shown, the output frequency can be varied from 60 Hz to 20 kHz by rotating potentiometer R2.

A portion of IC1's output voltage is fed to its noninverting input at pin 3. The voltage serves as a reference for capacitor C1, which is connected to the noninverting input at pin 2 of the IC. That capacitor continually charges and discharges around the reference voltage, and the result is a square-wave output. Capacitor C2 decouples the output.

2

Crystal Oscillators

The sources of the following circuits are contained in the Sources section, which begins on page 173. The figure number contained in the box of each circuit correlates to the source entry in the Sources section.

Fundamental-Frequency Crystal Oscillator
Easy Start-Up Crystal Oscillator
Crystal-Controlled Transistor Oscillator
Pierce Harmonic Oscillator (20 MHz)
Butler Emitter-Follower Oscillator (20 MHz)
Colpitts Oscillator
Low-Power, 5-V Driven Temperature-Compensated
  Crystal Oscillator (TXCO)
Crystal-Controlled Local Oscillator for SSB
  Transmitters
Crystal-Controlled Oscillator
Pierce Oscillator
CMOS Crystal Oscillator
Temperature-Compensated Crystal Oscillator
Crystal Oscillator
100-kHz Crystal Calibrator
Overtone Crystal Oscillator
20-MHz VHF Crystal Oscillator

Marker Generator
100-MHz VHF Crystal Oscillator
Two-Gate Quartz Oscillator
Crystal-Controlled Reflection Oscillator
Temperature-Compensated Crystal Oscillator
Overtone Crystal Oscillator
High-Frequency Crystal Oscillator
96-MHz Crystal Oscillator
Simple TTL Crystal Oscillator
Butler Emitter-Follower Oscillator (100 MHz)
Colpitts Harmonic Oscillator (Basic Circuit)
Colpitts Harmonic Oscillator (100 MHz)
International Crystal OF-1 Lo Oscillator
Crystal Oscillator
Overtone Crystal Oscillator
Butler Aperiodic Oscillator
Parallel-Mode Aperiodic Crystal Oscillator
Voltage-Controlled Crystal Oscillator

High-Frequency Crystal Oscillator

Crystal-Controlled Oscillator Operates from One
Mercury Cell

Colpitts Oscillator

Crystal-Controlled Oscillator

Low-Frequency Crystal Oscillator (10 to 150 kHz)

Overtone Crystal Oscillator

IC-Compatible Crystal Oscillator

Crystal Oscillator Provides Low Noise

Discrete Sequence Oscillator

1-MHz Pierce Oscillators

Simple CMOS Crystal Oscillator

Crystal Timebase

Low-Frequency Pierce Oscillator

1-MHz FET Crystal Oscillator

Pierce Crystal Oscillator

Overtone Oscillator with Crystal Switching

Crystal Oscillator

Fifth-Overtone Oscillator

Crystal-Controlled Butler Oscillator

Schmitt Trigger Crystal Oscillator

Overtone Oscillator (50 to 100 MHz)

Precision Clock Generator

Miller Oscillator (Crystal Controlled)

Butler Emitter-Follower Oscillator (Basic Circuit)

Butler Common-Base Oscillator (Basic Circuit)

CMOS Oscillator (1 to 4 MHz)

Pierce Harmonic Oscillator (100 MHz)

International Crystal OF-1 Hi Oscillator

Standard Crystal Oscillator for 1 MHz

TTL-Compatible Crystal Oscillator

Third-Overtone Crystal Oscillator

Crystal Checker

Crystal Oscillator/Doubler

Low-Frequency Crystal Oscillator

Varactor-Tuned 10-MHz Ceramic Resonator
Oscillator

10-MHz Crystal Oscillator

Pierce Harmonic Oscillator (Basic Circuit)

Tube-Type Crystal Oscillator

Crystal-Controlled Sine-Wave Oscillator

Crystal Oscillator

Stable Low-Frequency Crystal Oscillator

JFET Pierce Crystal Oscillator

Crystal Oscillator

Crystal-Controlled Signal Source

High-Frequency Signal Generator

Crystal-Stabilized IC Timer Can Provide
Subharmonic Frequencies

TTL Oscillator for 1 to 10 MHz

50-MHz VHF Crystal Oscillator

FUNDAMENTAL-FREQUENCY CRYSTAL OSCILLATOR

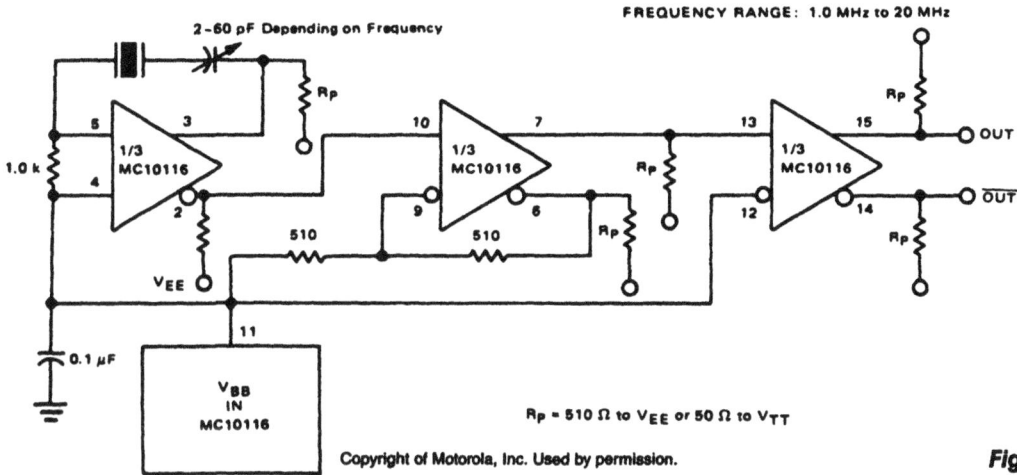

FREQUENCY RANGE: 1.0 MHz to 20 MHz

2-60 pF Depending on Frequency

R_p = 510 Ω to V_{EE} or 50 Ω to V_{TT}

Copyright of Motorola, Inc. Used by permission.

Fig. 2-1

For frequencies below 20 MHz, a fundamental-frequency crystal can be used and the resonant tank is no longer required. Also, at this lower frequency range the typical MECL 10,000 propagation delay of 2 ns becomes small compared to the period of oscillation, and it becomes necessary to use a noninverting output. Thus, the MC10116 oscillator section functions simply as an amplifier. The 1.0 kΩ resistor biases the line receiver near V_{BB} and the 0.1-μF, capacitor is a filter capacitor for the V_{BB} supply. The capacitor, in series with the crystal, provides for minor frequency adjustments. The second section of the MC10116 is connected as a Schmitt-trigger circuit; this ensures good MECL edges from a rather slow, less than 20-MHz input signal. The third stage of the MC10116 is used as a buffer and to give complementary outputs from the crystal oscillator circuit. The circuit has a maximum operating frequency of approximately 20 MHz and a minimum of approximately 1 MHz; it is intended for use with a crystal which operates in the fundamental mode of oscillation.

EASY START-UP CRYSTAL OSCILLATOR

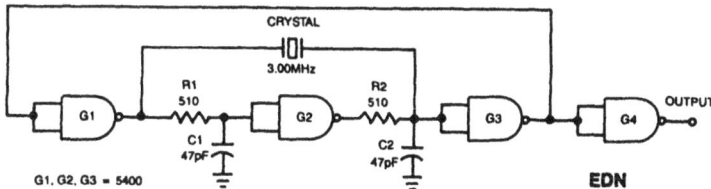

CRYSTAL
3.00MHz

R1 510   R2 510   OUTPUT

C1 47pF   C2 47pF

G1, G2, G3 = 5400

EDN

Fig. 2-2

This low cost, crystal-controlled oscillator uses one TTL gate. Two factors ensure oscillator start-up: The connection of NAND gates G1, G2, and G3 into an unstable logic configuration and the high loop gain of the three inverters. Values of R1, R2, C1, and C2 aren't critical; select them so the oscillator operates at a frequency 70 to 90% higher than the crystal frequency when the crystal is disconnected. For 1−2 MHz operation, a low-power 54L00 IC is recommended; for a 2−6 MHz, a standard 5400 type; and for 6−20 MHz, a 54H00 or 54S00.

CRYSTAL-CONTROLLED TRANSISTOR OSCILLATOR

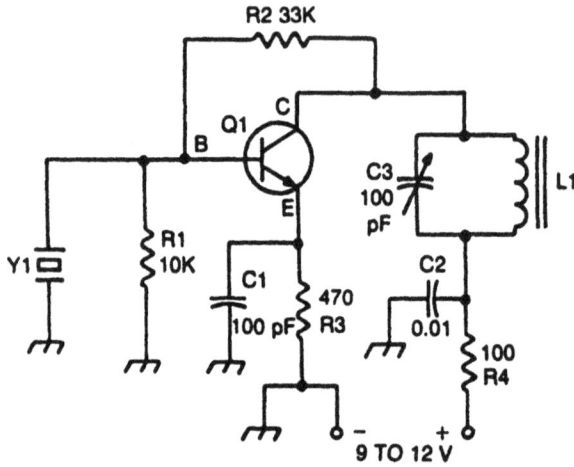

Fig. 2-3

PIERCE HARMONIC OSCILLATOR (20 MHz)

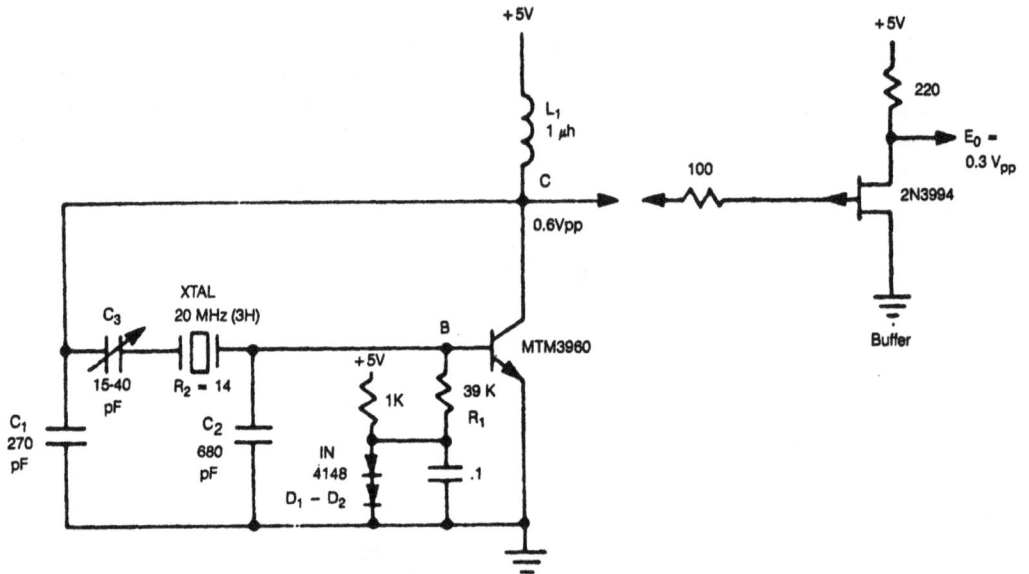

Fig. 2-4

JOHN WILEY & SONS

This circuit has excellent short-term frequency stability because the external load tied across the crystal is mostly capacitive rather than resistive, giving the crystal a high in-circuit Q.

BUTLER EMITTER-FOLLOWER OSCILLATOR (20 MHz)

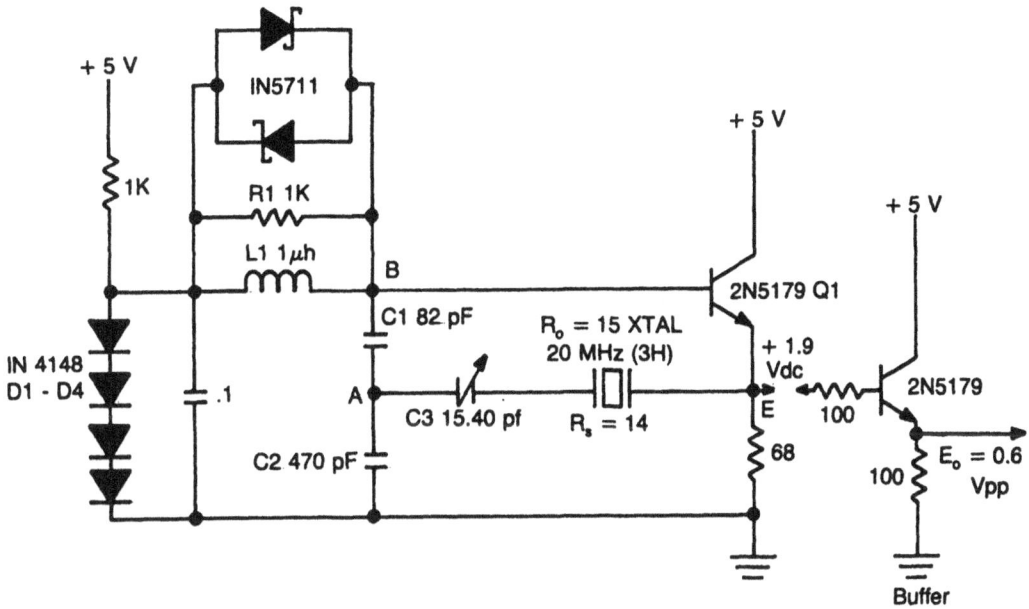

Fig. 2-5

JOHN WILEY & SONS

COLPITTS OSCILLATOR

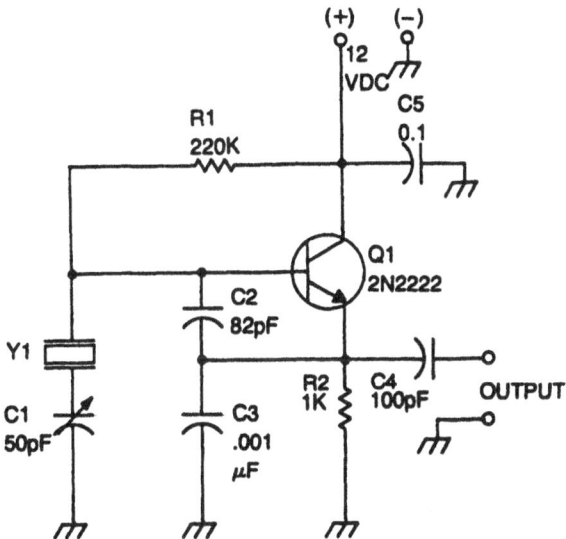

This circuit will operate with fundamental-mode crystals in the range of 1 to 20 MHz. Feedback is controlled by capacitor voltage divider C2/C3. The rf voltage across the emitter resistor provides the basic feedback signal.

TAB BOOKS

Fig. 2-6

LOW-POWER 5-V DRIVEN TEMPERATURE-COMPENSATED CRYSTAL OSCILLATOR (TXCO)

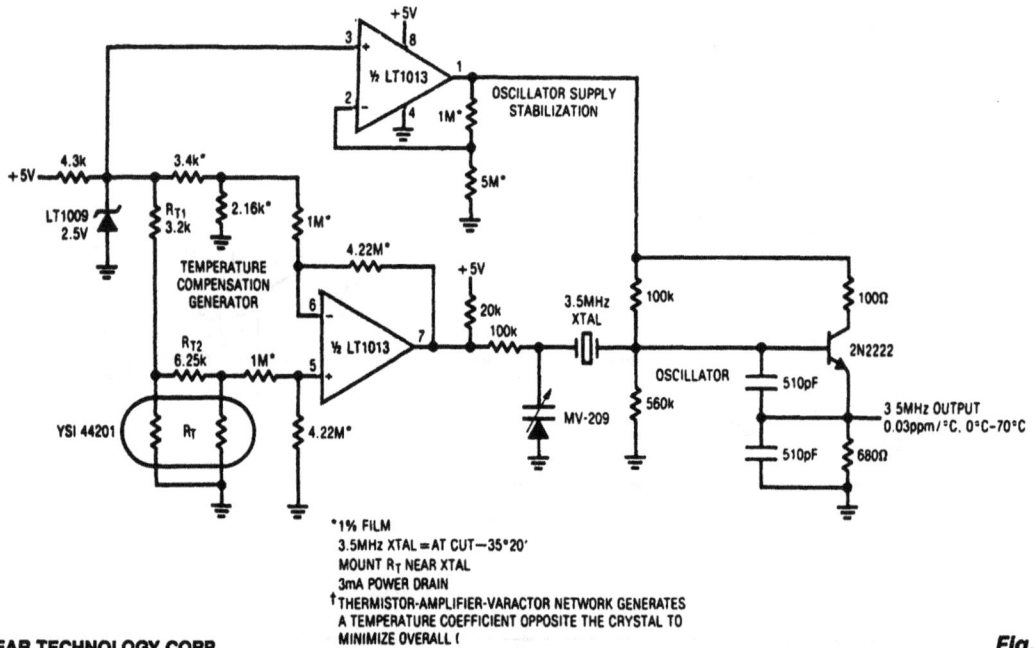

*1% FILM
3.5MHz XTAL = AT CUT—35°20'
MOUNT R_T NEAR XTAL
3mA POWER DRAIN
†THERMISTOR-AMPLIFIER-VARACTOR NETWORK GENERATES
A TEMPERATURE COEFFICIENT OPPOSITE THE CRYSTAL TO
MINIMIZE OVERALL (

LINEAR TECHNOLOGY CORP.

Fig. 2-7

CRYSTAL-CONTROLLED LOCAL OSCILLATOR FOR SSB TRANSMITTERS

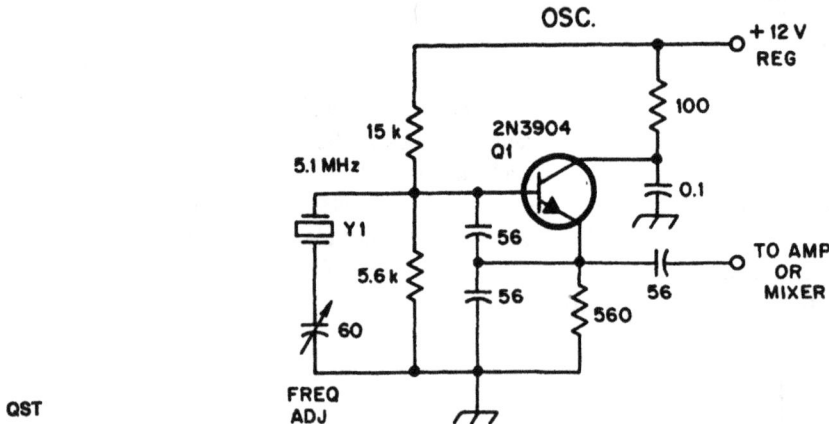

QST

Fig. 2-8

This oscillator can contain several switched crystals to provide channelized operation. A buffer amplifier can be added, if desired.

CRYSTAL-CONTROLLED OSCILLATOR

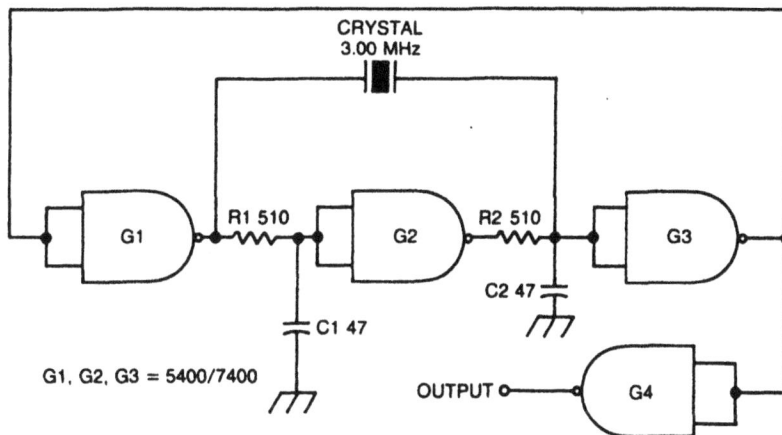

HAM RADIO

Fig. 2-9

This circuit oscillates without the crystal. With the crystal in the circuit, the frequency will be that of the crystal. The circuit has good starting characteristics—even with the poorest crystals.

PIERCE OSCILLATOR

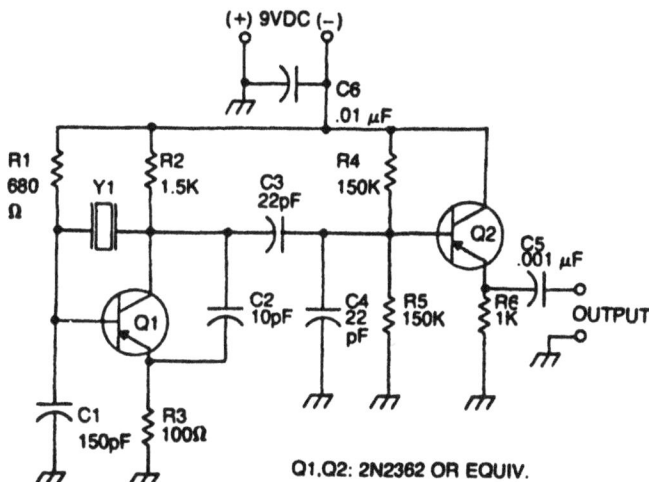

TAB BOOKS

Fig. 2-10

The oscillator transistor is Q1, and the crystal is placed between the collector and base. Feedback is improved by the use of the collector-emitter capacitor C2. Transistor Q2 is used as an output buffer.

31

CMOS CRYSTAL OSCILLATOR

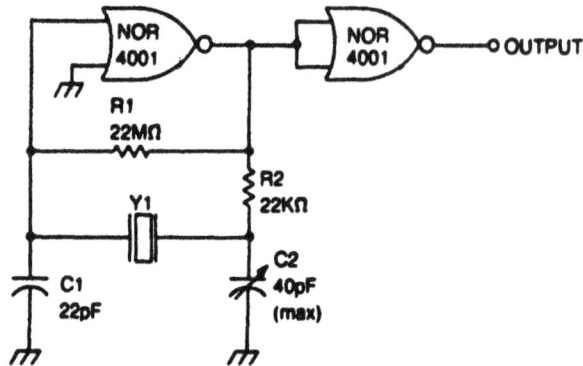

Fig. 2-11

TAB BOOKS

This circuit has a frequency range of 0.5 to 2.0 MHz. Frequency can be adjusted to a precise value with trimmer capacitor C2. The second NOR gate serves as an output buffer.

TEMPERATURE-COMPENSATED CRYSTAL OSCILLATOR

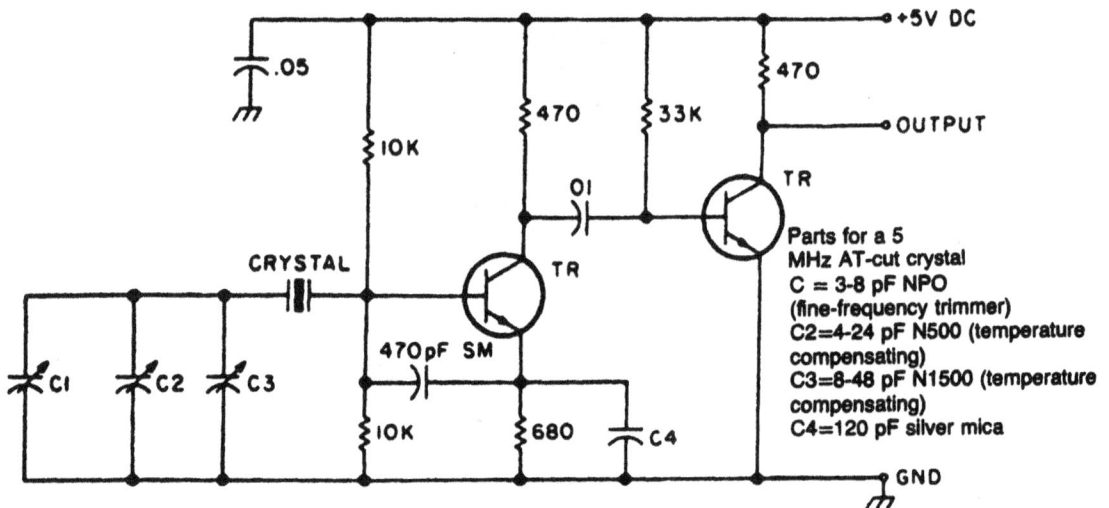

Parts for a 5
MHz AT-cut crystal
C = 3-8 pF NPO
(fine-frequency trimmer)
C2=4-24 pF N500 (temperature compensating)
C3=8-48 pF N1500 (temperature compensating)
C4=120 pF silver mica

73 AMATEUR RADIO

Fig. 2-12

Two different negative-coefficient capacitors are blended to produce the desired change in capacitance to counteract or compensate for the decrease in frequency of the "normal" AT-cut characteristics.

CRYSTAL OSCILLATOR

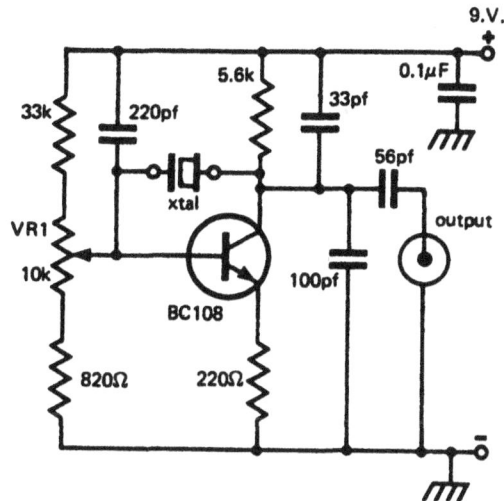

ELECTRONICS TODAY INTERNATIONAL

Fig. 2-13

This circuit provides reliable oscillation and an output close to one volt peak-to-peak. Power consumption is around 1 mA from a 9-V supply.

100-kHz CRYSTAL CALIBRATOR

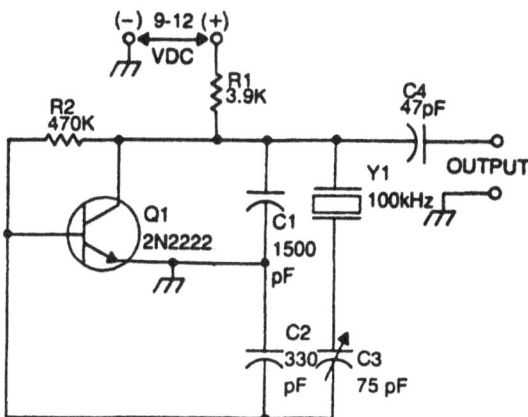

This circuit is often used by amateur radio operations, shortwave listeners, and other operators of shortwave receivers to calibrate the dial pointer. The oscillator operates at a fundamental frequency of 100 kHz, and the harmonics are used to locate points on the shortwave dial, provided that the output of the calibrator is coupled to the antenna circuit of the receiver. The crystal shunts the feedback-voltage divider, and is in series with a variable capacitor (C3) that is used to set the actual operating frequency of the calibrator.

TAB BOOKS

Fig. 2-14

OVERTONE CRYSTAL OSCILLATOR

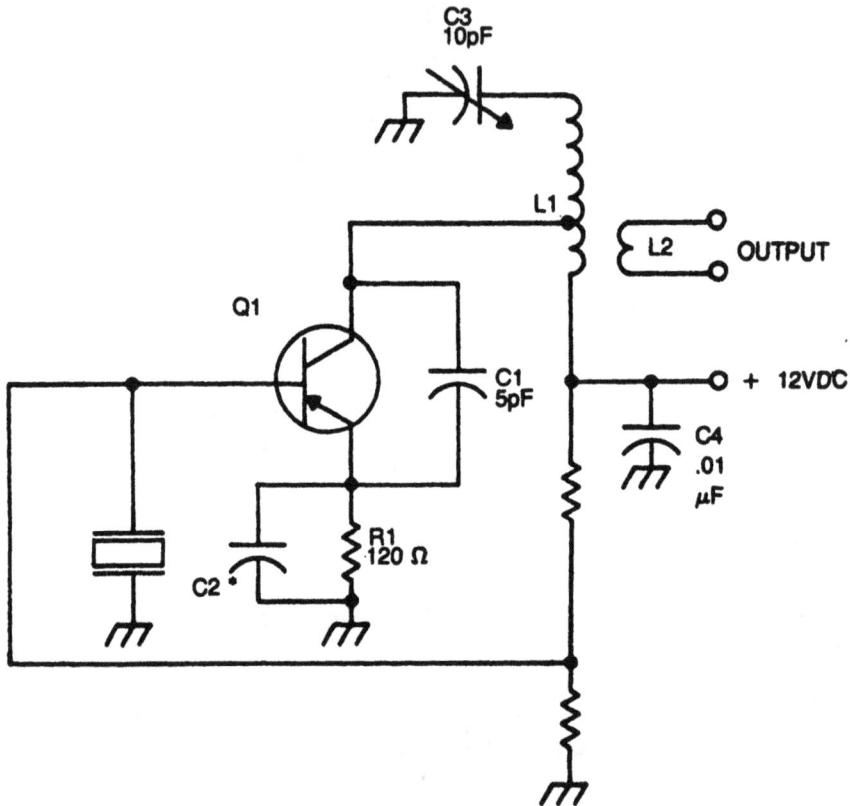

Fig. 2-15

The crystal element in this circuit is connected directly between the base and ground. Capacitor C1 is used to improve the feedback due to the internal capacitances of the transistor. This capacitor should be mounted as close as possible to the case of the transistor. The LC tank circuit in the collector of the transistor is tuned to the overtone frequency of the crystal. The emitter resistor capacitor must have a capacitive reactance of approximately 90-Ω at the frequency of operation. The tap on inductor L1 is used to match the impedance of the collector of the transistor. In most cases, the optimum placement of this tap is approximately one-third from the cold end of the coil. The placement of this tap is a trade-off between stability and maximum power output. The output signal is taken from a link coupling coil, L2, and operates by transformer action.

20-MHz VHF CRYSTAL OSCILLATOR

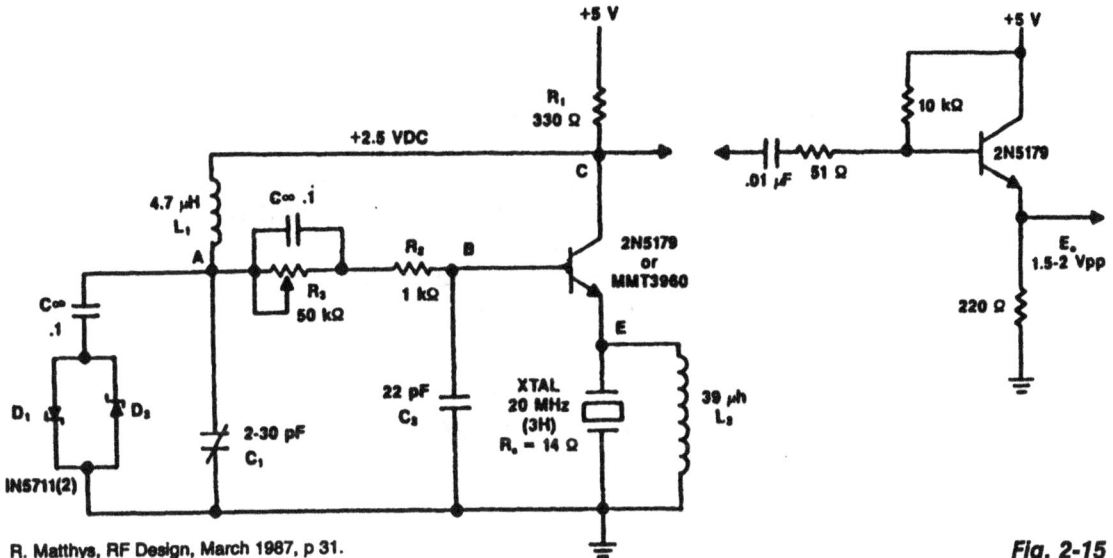

R. Matthys, RF Design, March 1987, p 31.

Fig. 2-15

A typical circuit at 20 MHz is shown. The crystal, which has an internal series resistance R_s of 14 Ω, oscillates at its third harmonic. The diode clamp D1 and D2 provides a constant amplitude control. The transistor operates continuously in a linear mode over a complete cycle of oscillation, and reflects a reasonably constant load across the crystal at all times.

MARKER GENERATOR

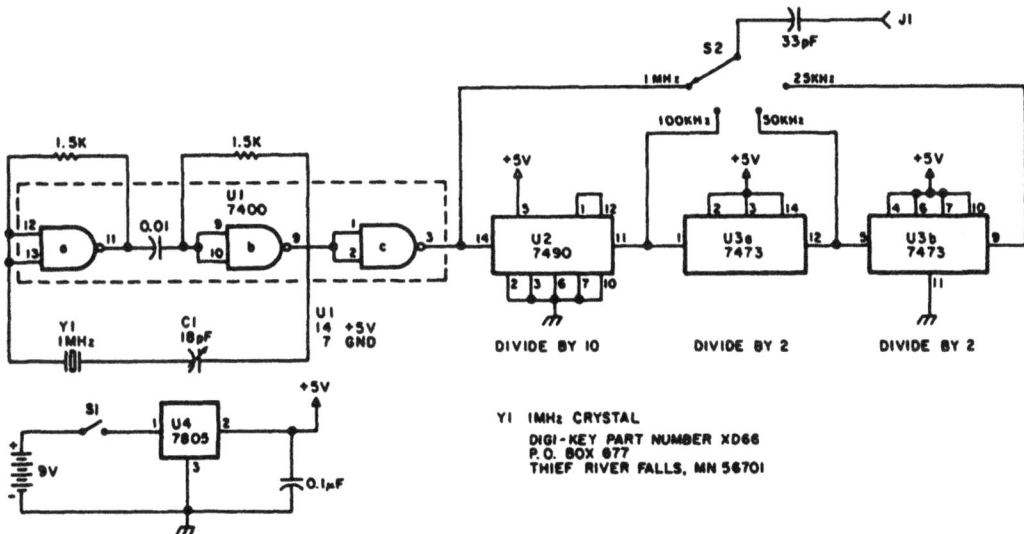

YI 1MHz CRYSTAL
DIGI-KEY PART NUMBER XO66
P.O. BOX 677
THIEF RIVER FALLS, MN 56701

73 AMATEUR RADIO

Fig. 2-17

35

MARKER GENERATOR (continued)

The oscillator section uses three sections of a 7400 quad NAND gate integrated circuit. The 1-MHz signal from the oscillator is fed into a 7490 decade counter configured to divide by ten, providing the 100-kHz signal. To obtain the 50 and 25 kHz outputs, the 100-kHz signal is further divided by 7473 dual J-K flip-flop. The first half of the 7473 divides the 100-kHz signal by two, yielding the 50 kHz signal. The second half of the 7473 again divides by two, giving the 25 kHz signal. S2 selects the output, a square wave, rich in harmonics. The generator can be powered from any convenient 6 to 12 Vdc source. A 7805 fixed-voltage regulator supplies the regulated voltage for the oscillator and the divider chips. The generator described here is powered by a 9-V transistor radio battery.

100-MHz VHF CRYSTAL OSCILLATOR

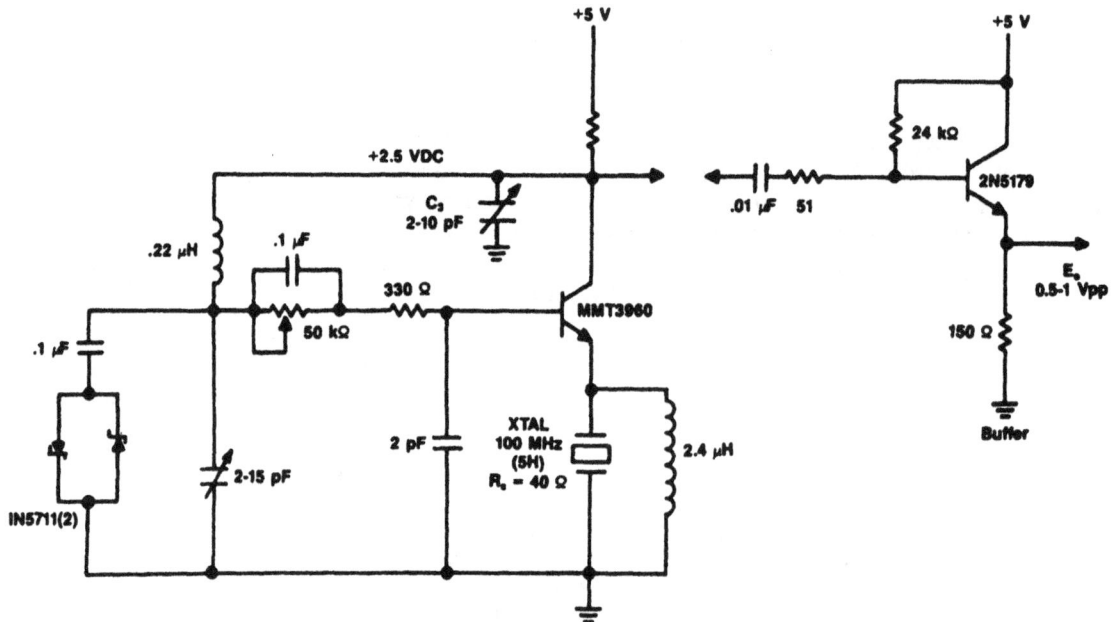

R. Matthys, RF Design, March 1987, p. 31.

Fig. 2-18

Figure 2-18 shows a 100-MHz oscillator operating on the fifth harmonic. Again to maintain the transistor's gain, note the increase in the collector's load resistance R1 because of the increase in the quartz crystal's internal series resistance R_s. C3 is needed at frequencies above 50 MHz to tune out the shunting effect of L1 on R1, to maintain a high load resistance for the transistor and get enough gain for oscillation. The equivalent series $R_L C_L$, load across the crystal is 8.2 Ω (R_L) and 200 pF (C_L).

TWO-GATE QUARTZ OSCILLATOR

A SN7400 quartz crystal and a resistor provide a square-wave output of approximately 3.5 V. The circuit operates reliably at frequencies from 120 kHz to 4 MHz.

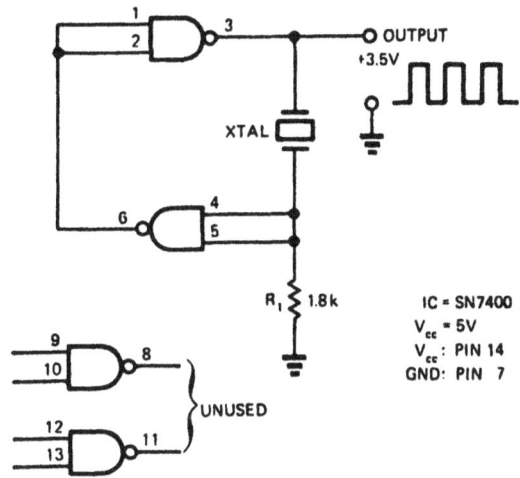

OUTPUT
+3.5V

XTAL

R_1 1.8k

IC = SN7400
V_{cc} = 5V
V_{cc} : PIN 14
GND: PIN 7

UNUSED

EDN

Fig. 2-19

CRYSTAL-CONTROLLED REFLECTION OSCILLATOR

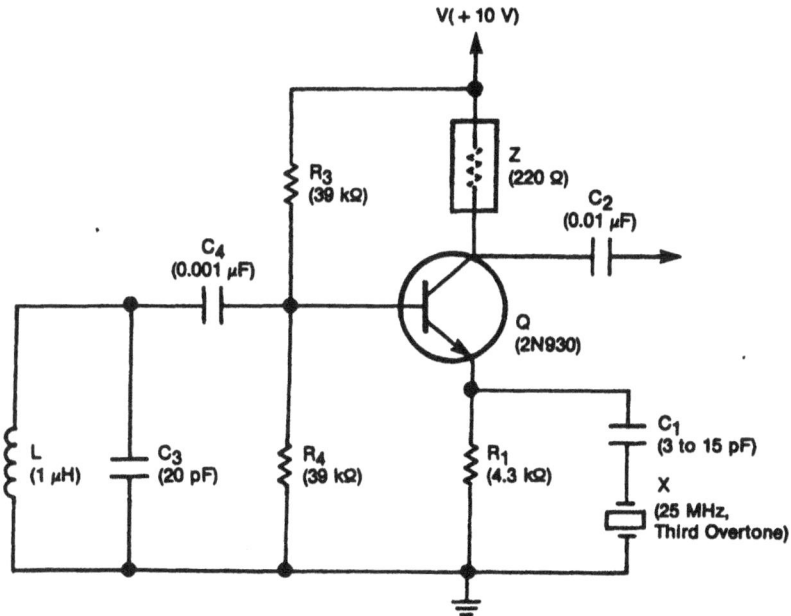

V(+ 10 V)

R_3 (39 kΩ)

Z (220 Ω)

C_2 (0.01 μF)

C_4 (0.001 μF)

Q (2N930)

L (1 μH)

C_3 (20 pF)

R_4 (39 kΩ)

R_1 (4.3 kΩ)

C_1 (3 to 15 pF)

X (25 MHz, Third Overtone)

NASA

Fig. 2-20

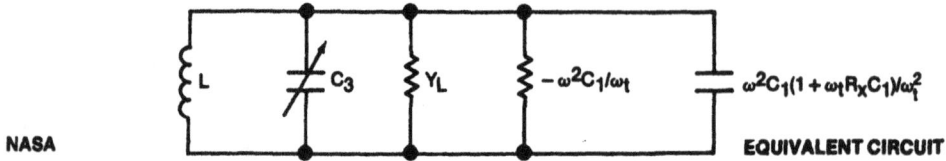

NASA

$$-\omega^2 C_1/\omega_t \qquad \omega^2 C_1(1 + \omega_t R_X C_1)/\omega_t^2$$

EQUIVALENT CIRCUIT

Fig. 2-20

This unit is easily tunable and stable, consumes little power, and costs less than other types of oscillators that operate at the same frequencies. This unusual combination of features is made possible by a design concept that includes operation of the transistor well beyond the 3 dB frequency of its current-versus-frequency curve. The concept takes advantage of newly available crystals that resonate at frequencies up to about 1 GHz.

The emitter of transistor Q is connected with variable capacitor C1 and series-resonant crystal X. The emitter is also connected to ground through bias resistor R1. The base is connected to the parallel combination of inductor L and capacitor C3 through dc-blocking capacitor and C4 and is forward biased with respect to the emitter by resistors R3 and R4. Impedance Z could be the 220-Ω resistor shown or any small impedance that enables the extraction of the output signal through coupling capacitor C2. If Z is a tuned circuit, it is tuned to the frequency of the crystal.

TEMPERATURE-COMPENSATED CRYSTAL OSCILLATOR

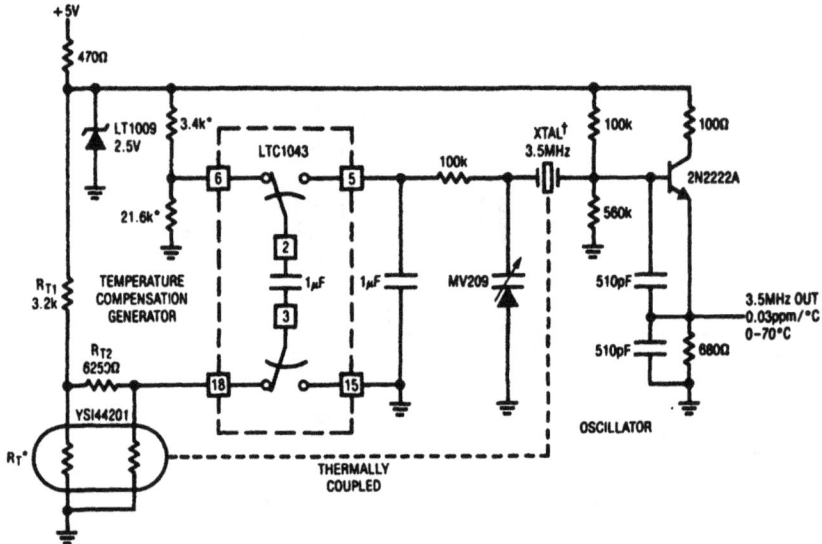

LINEAR TECHNOLOGY CORP.

Fig. 2-21

This circuit uses LTC1043 to differentiate between a temperature sensing network and a dc reference. The single-ended output biases a varactor-tuned crystal oscillator to compensate drift. The varactor crystal network has high dc impedance, eliminating the need for an LTC1043 output amplifier.

OVERTONE CRYSTAL OSCILLATOR

FREQUENCY RANGE:

20 MHz to 100 MHz, Dependent on
Crystal Frequency and Tank Tuning

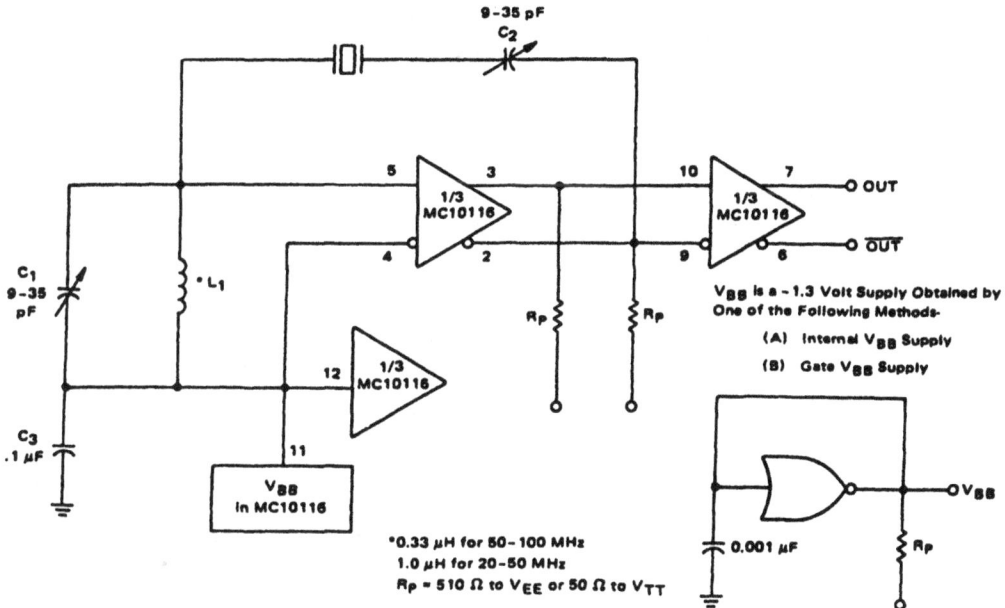

V_{BB} is a ~1.3 Volt Supply Obtained by
One of the Following Methods-

(A) Internal V_{BB} Supply

(B) Gate V_{BB} Supply

*0.33 μH for 50-100 MHz
1.0 μH for 20-50 MHz
R_p = 510 Ω to V_{EE} or 50 Ω to V_{TT}

Fig. 2-22

This circuit employs an adjustable resonant tank circuit that ensures operation at the desired crystal overtone. C1 and L1 form the resonant tank circuit, which with the values specified as a resonant frequency, are adjustable from approximately 50 to 100 MHz. Overtone operation is accomplished by adjusting the tank circuit frequency at or near the desired frequency. The tank circuit exhibits a low impedance shunt to off-frequency oscillations and a high impedance to the desired frequency, which allows feedback from the output. Operation in this manner guarantees that the oscillator will always start at the correct overtone.

HIGH-FREQUENCY CRYSTAL OSCILLATOR

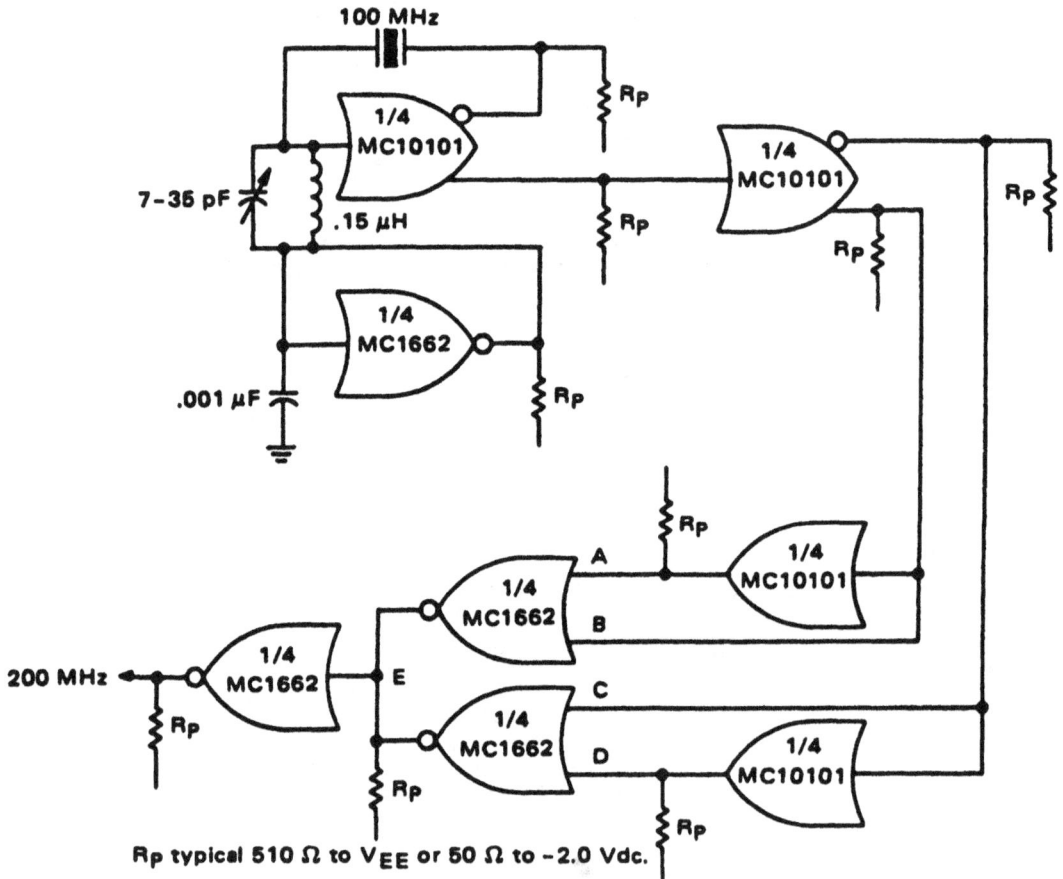

MOTOROLA

Fig. 2-23

One section of the MC10101 is connected as a 100-MHz crystal oscillator with the crystal in series with the feedback loop. The LC tank circuit tunes the 100-MHz harmonic of the crystal and can be used to calibrate the circuit to the exact frequency. A second section of the MC10101 buffers the crystal oscillator and gives complementary 100-MHz signals. The frequency doubler consists of two MC10101 gates as phase shifters and two MC1662 NOR gates. For a 50% duty cycle at the output, the delay to the true and complement 100-MHz signals should be 90°. This can be built precisely with 2.5-ns delay lines for the 200-MHz output or approximated by the two MC10101 gates as shown.

96-MHz CRYSTAL OSCILLATOR

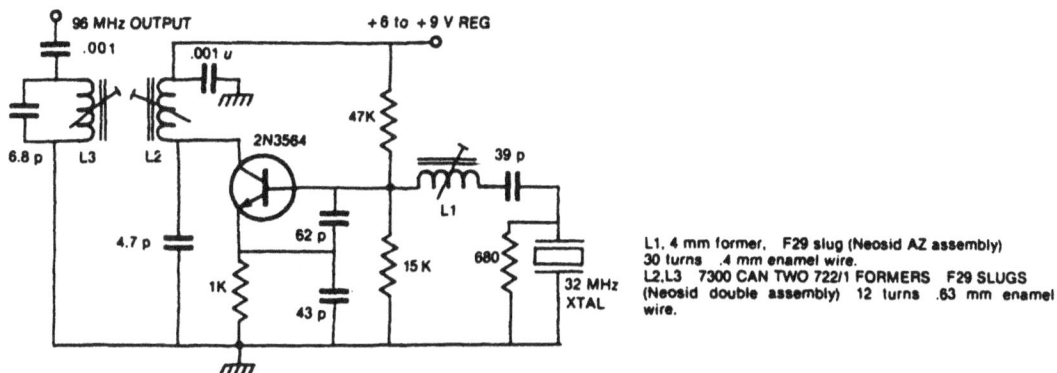

L1, 4 mm former, F29 slug (Neosid AZ assembly)
30 turns .4 mm enamel wire.
L2,L3 7300 CAN TWO 722/1 FORMERS F29 SLUGS
(Neosid double assembly) 12 turns .63 mm enamel
wire.

ELECTRONICS TODAY INTERNATIONAL   *Fig. 2-24*

By using a crystal between 27.5 and 33 MHz, the 3rd harmonic will deliver between 82.5 and 99 MHz.

SIMPLE TTL CRYSTAL OSCILLATOR

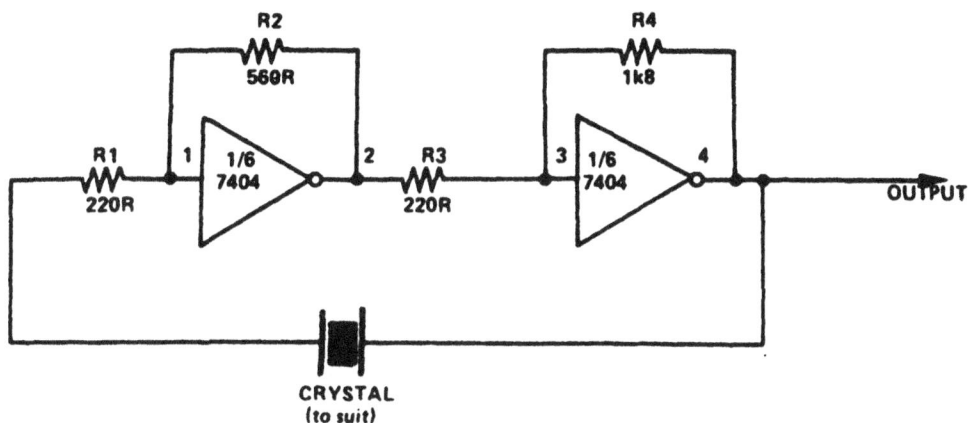

ELECTRONICS TODAY INTERNATIONAL   *Fig. 2-25*

This simple and cheap crystal oscillator compasses one third of a 7404, four resistors and a crystal. The inverters are biased into their linear regions by R1 to R4, and the crystal provides the feedback. Oscillation can only occur at the crystal's fundamental frequency.

41

BUTLER EMITTER-FOLLOWER OSCILLATOR (100 MHz)

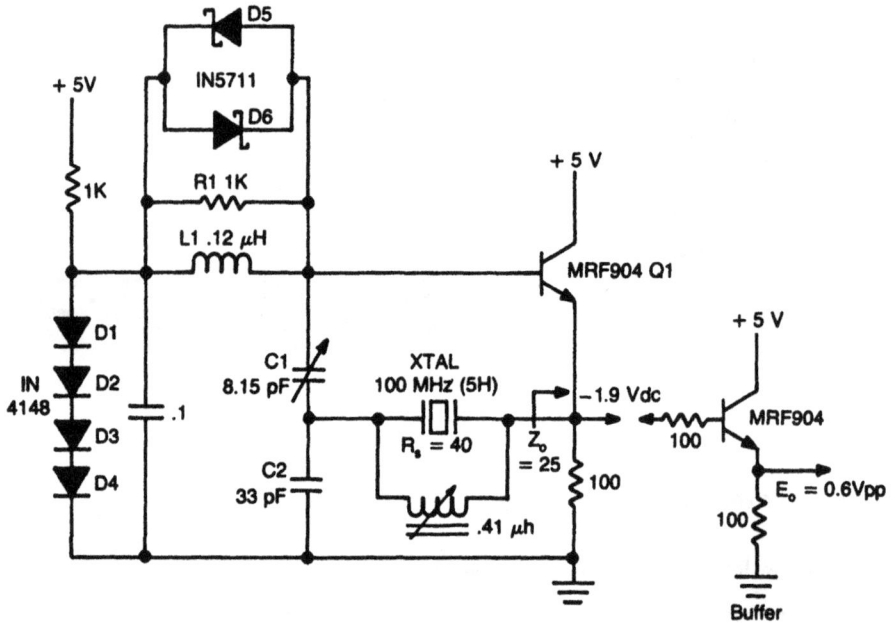

Fig. 2-26

JOHN WILEY & SONS

This circuit has good performance without any parasitics because emitter-follower amplifier has a gain of only one with built-in negative feedback to stabilize its gain.

COLPITTS HARMONIC OSCILLATOR (BASIC CIRCUIT)

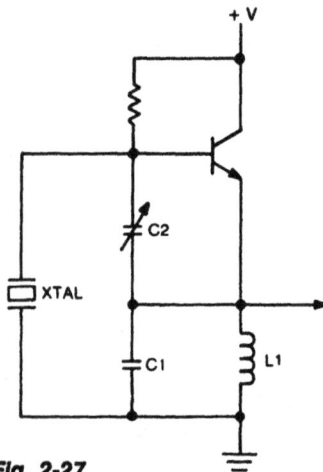

This circuit operates 30–200 ppm above series resonance. Physically simple, but analytically complex. It is inexpensive and has fair frequency stability.

Fig. 2-27

JOHN WILEY & SONS

COLPITTS HARMONIC OSCILLATOR (100 MHz)

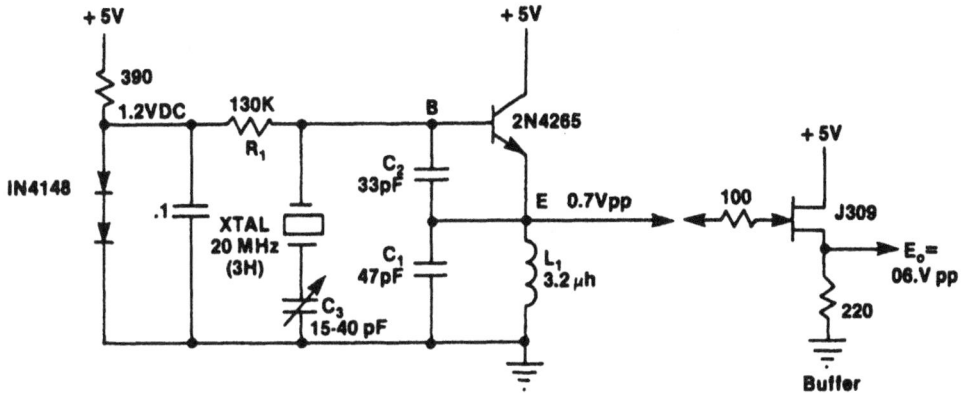

JOHN WILEY & SONS

Fig. 2-28

L1C1 are selected to be resonant at a frequency below the desired crystal harmonic but above the crystal's next lower odd harmonic. C2 should have a value of 30 – 70 pF, independent of the oscillation frequency. There is no requirement for any specific ratio of C1/C2, but practical harmonic circuits seem to work best when C1 is approximately 1 to 3 times the value of C2. Diodes D1, D2, D3 provide a simple regulated bias supply. The resistance of R1 should be as high as possible, because it affects the crystal's in-circuit Q.

INTERNATIONAL CRYSTAL OF-1 LO OSCILLATOR

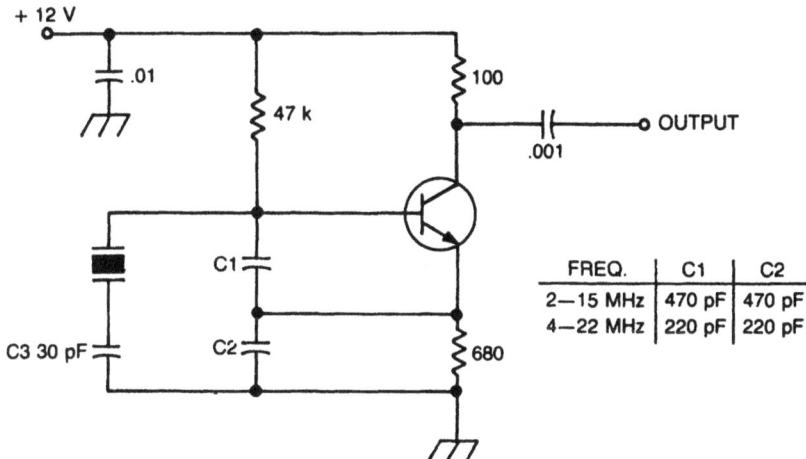

FREQ.	C1	C2
2—15 MHz	470 pF	470 pF
4—22 MHz	220 pF	220 pF

HAM RADIO

Fig. 2-29

International Crystal OF-1 LO oscillator circuit for fundamental-node crystals.

CRYSTAL OSCILLATOR

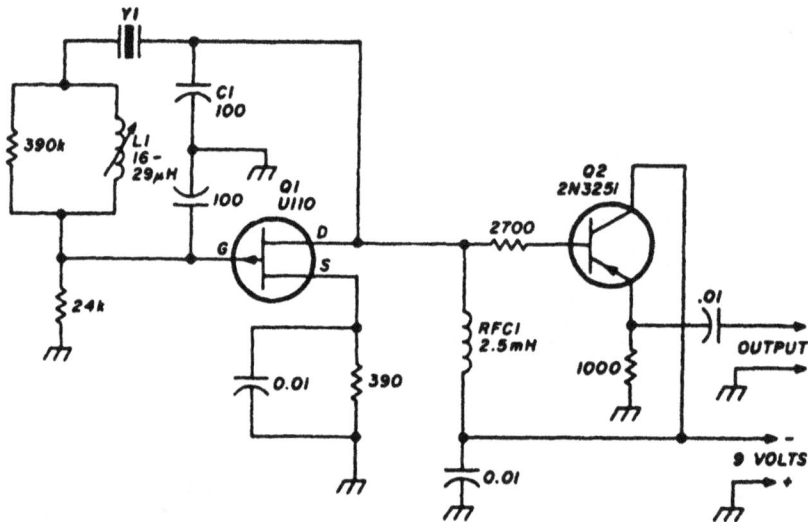

Fig. 2-30

This stable VXO using 6- or 8-MHz crystals uses a capacitor and an inductor to achieve frequency pulling on either side of series resonance.

OVERTONE CRYSTAL OSCILLATOR

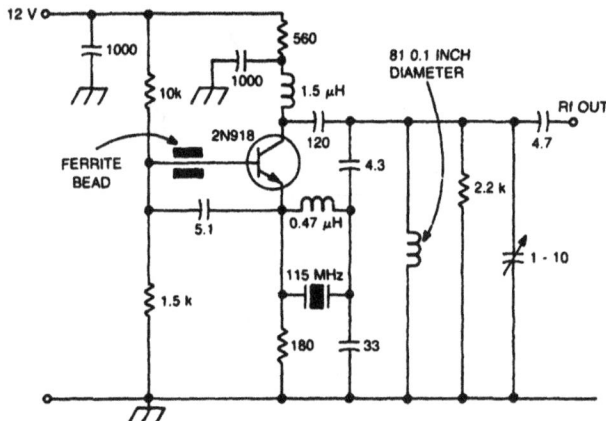

HAM RADIO

Fig. 2-31

This design is for high reliability over a wide temperature range using fifth and seventh overtone crystals. The inductor in parallel with the crystal causes antiresonance of crystal C_0 to minimize loading. This technique is commonly used with overtone crystals.

BUTLER APERIODIC OSCILLATOR

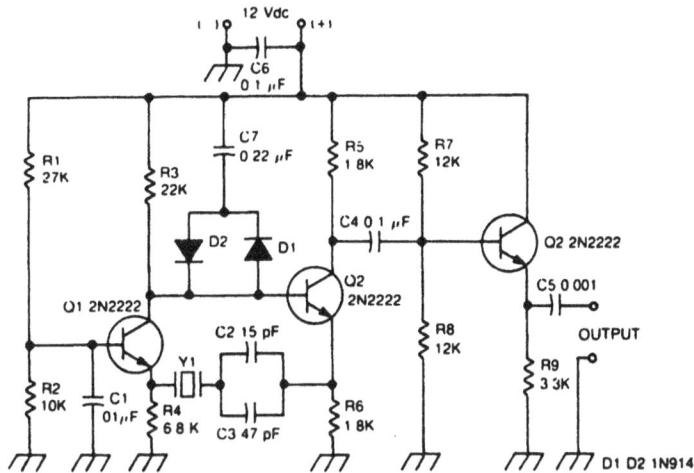

Fig. 2-32

This circuit works well in the range of 50 to 500 kHz. Slight component modifications are needed for higher frequency operation. For operation over 3000 kHz, select a transistor that provides moderate gain (in the 60 to 150 range) at the frequency of operation and a gain-bandwidth product of at least 100 MHz.

PARALLEL-MODE APERIODIC CRYSTAL OSCILLATOR

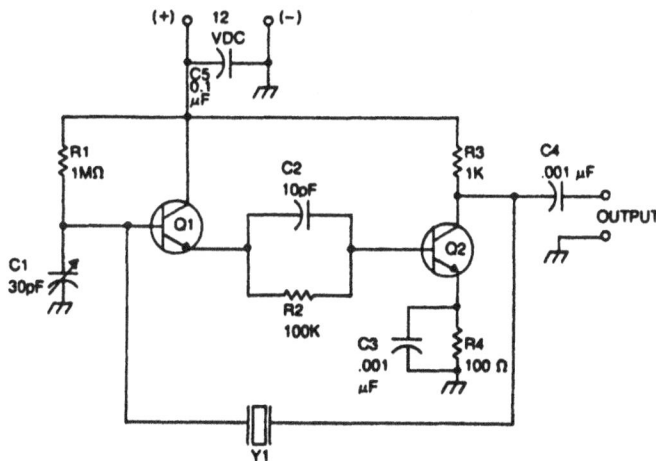

Fig. 2-33

The crystal is placed between the collector of the output stage and the base of the input stage. The frequency of oscillation can be set to a precise value with trimmer capacitor C1. The range of operation for this circuit is 500 kHz to 10 MHz. Extend the range downward (100 kHz) by increasing the value of C1 to 75 pF and by increasing the value of C2 to 22 pF.

VOLTAGE-CONTROLLED CRYSTAL OSCILLATOR

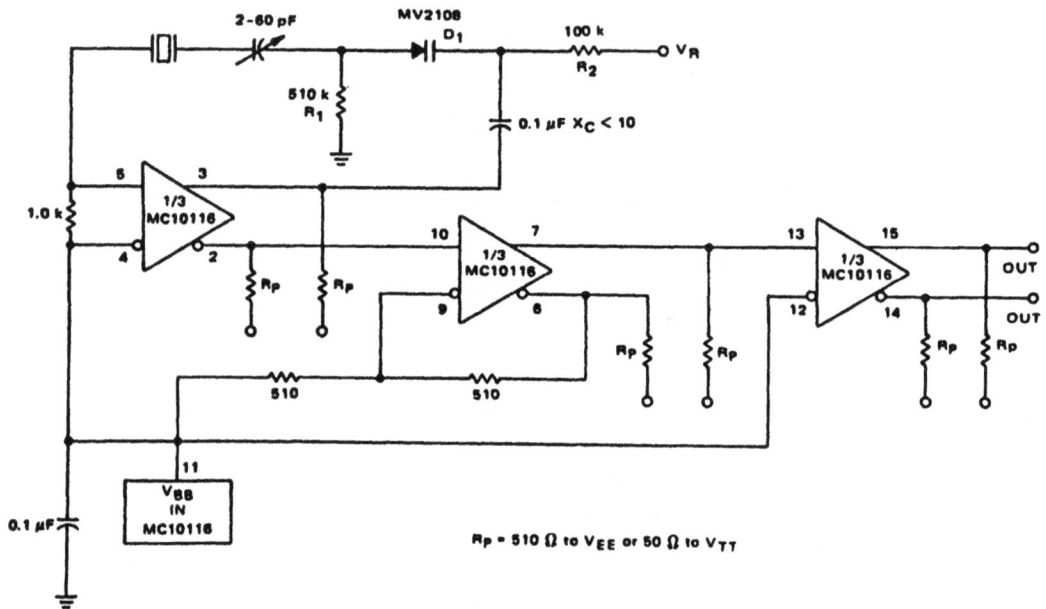

NOMINAL FREQUENCY	DEVIATION	
MHz	+PPM	–PPM
1.0000	57.0	48.0
1.8432	95.5	80.3
10.000	197.4	202.8
15.000	325.4	322.9

Fig. 2-34

A voltage-variable capacitance tuning diode is placed in series with the crystal feedback path. Changing the voltage on V_R varies the tuning diode capacitance and tunes the oscillator. The 510-kΩ resistor, R1, establishes a reference voltage for V_R—ground is used in this example. A 100-kΩ resistor, R2, isolates the tuning voltage from the feedback loop and 0.1-μF capacitor C2 provides ac coupling to the tuning diode. The circuit operates over a tuning range of 0 to 25 V. It is possible to change the tuning range from 0 to 25 V by reversing the tuning diode D1. Center frequency is set with the 2–60 pF trimmer capacitor. Deviation on either side of center is a function of the crystal frequency. The table in Fig. 2-34 shows measured deviation in parts per million for several tested crystals.

HIGH-FREQUENCY CRYSTAL OSCILLATOR

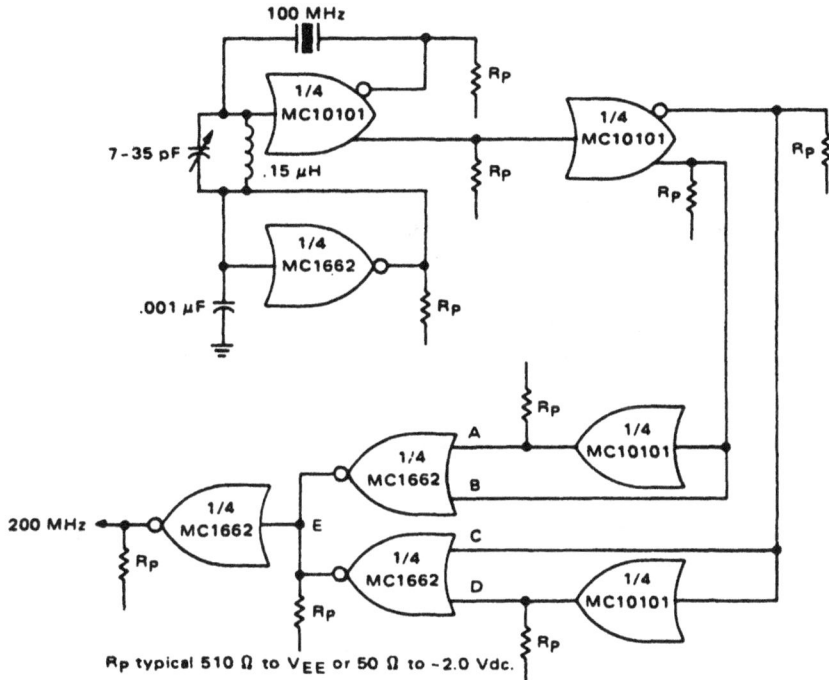

100 MHz

1/4 MC10101

Rp

7 - 35 pF

.15 µH

1/4 MC10101

Rp

Rp

Rp

.001 µF

1/4 MC1662

Rp

1/4 MC1662

A

B

Rp

1/4 MC10101

200 MHz

1/4 MC1662

E

Rp

1/4 MC1662

C

D

1/4 MC10101

Rp

Rp typical 510 Ω to V_{EE} or 50 Ω to −2.0 Vdc.

200 MHz Crystal Oscillator

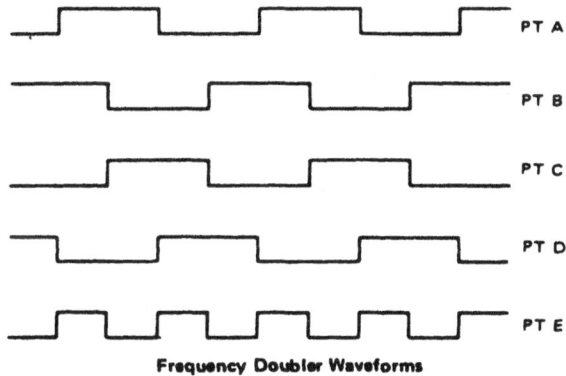

PT A

PT B

PT C

PT D

PT E

Frequency Doubler Waveforms

Fig. 2-35

A high-speed oscillator is possible by combining an MECL 10-kΩ crystal oscillator with an MECL III frequency doubler, as shown. One section of the MC10101 is connected as a 100-MHz crystal oscillator with the crystal in series with the feedback loop. The LC tank circuit tunes the 100-MHz harmonic of the crystal and can be used to calibrate the circuit to the exact frequency. A second section of the MC10101 buffers the crystal oscillator and gives complementary 100-MHz signals. The frequency doubler consists of two MC10101 gates as phase shifters and two MC1662 NOR gates. For a 50% duty cycle at the output, the delay to the true and complement 100-MHz signals should be 90°. This can be built precisely with 2.5-ns delay lines for the 200-MHz output or it can be approximated by the two MC10101 gates. The gates are easier to incorporate and cause only a slight skew in output signal duty cycle. The MC1662 gates combine the 4-phase 100-MHz signals, as shown in Fig. B. The outputs of the MC1662's are wire-OR connected to give the 200-MHz signal. MECL III gates are used because of the bandwidth required for 200-MHz signals. One of the remaining MC1662 gates is used as a V_{BB} bias generator for the oscillator. By connecting the NOR output to the input, the circuit stays in the center of the logic swing or at V_{BB}. A 0.001-μF capacitor ensures that the V_{BB} circuit does not oscillate.

CRYSTAL-CONTROLLED OSCILLATOR OPERATES FROM ONE MERCURY CELL

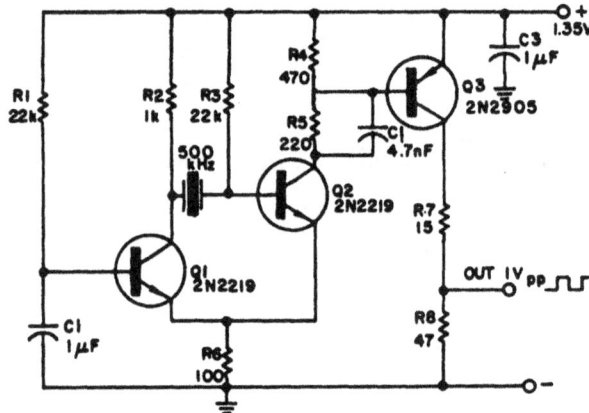

Inexpensive crystal controlled oscillator operates from a 1.35-volt source.

ELECTRONIC DESIGN

Fig. 2-36

The circuit is powered by a single 1.35-V mercury cell and it provides a 1-V square-wave output. As shown, the crystal is a tuned circuit between transistors Q1 and Q2, which are connected in the common-emitter configuration. Positive feedback, provided by means of R, permits oscillation. The signal at the collector of Q2 is squared by Q3, which switches between cutoff and saturation. R7 permits short-circuit-proof operation.

COLPITTS OSCILLATOR

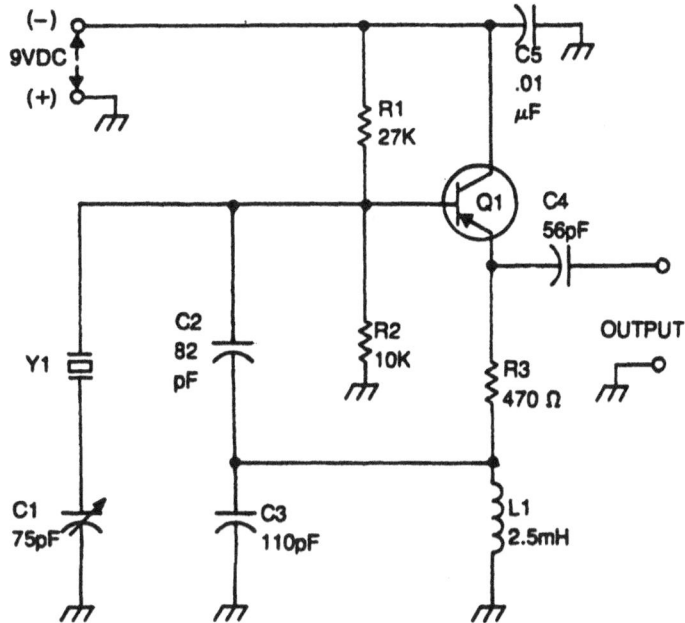

TAB BOOKS

Fig. 2-37

Bias for the pnp bipolar transistor is provided by resistor voltage-divider network R1/R2. The collector of the oscillator transistor is kept at ac ground by capacitor C5, placed close to the transistor. Feedback is provided by capacitor voltage divider C2/C3.

CRYSTAL-CONTROLLED OSCILLATOR

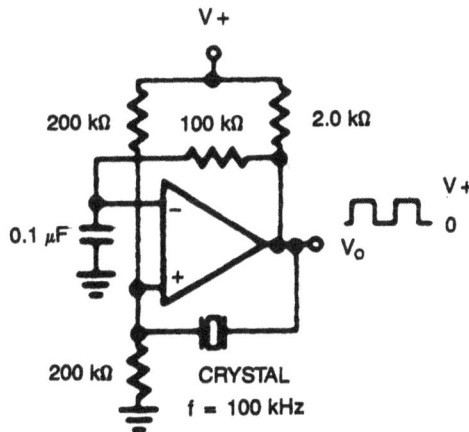

SIGNETICS

Fig. 2-38

LOW-FREQUENCY CRYSTAL OSCILLATOR—10 to 150 kHz

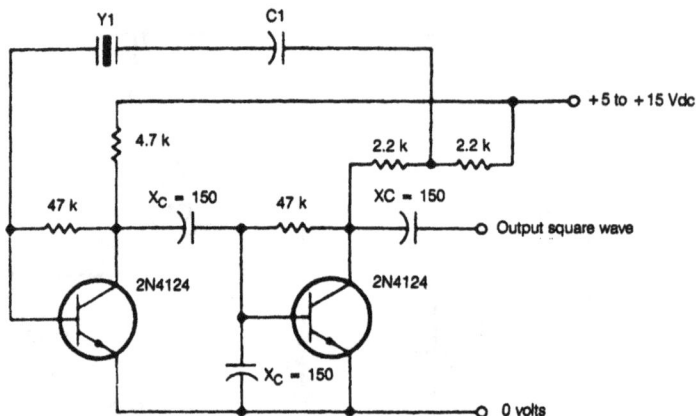

HAM RADIO

Fig. 2-39

C1 in series with the crystal, can be used to adjust the oscillator output frequency. The value can range between 20 pF and 0.01 μF, or it can be a trimmer capacitor and that will approximately equal the crystal load capacitance. X values are approximate and can vary for most circuits and frequencies; this is also true for resistance values. Adequate power supply decoupling is required; local decoupling capacitors near the oscillator are recommended. All leads should be extremely short in high-frequency circuits.

OVERTONE CRYSTAL OSCILLATOR

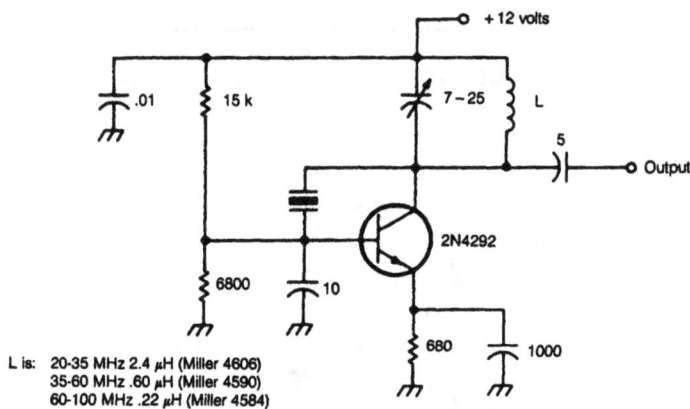

L is: 20-35 MHz 2.4 μH (Miller 4606)
       35-60 MHz .60 μH (Miller 4590)
       60-100 MHz .22 μH (Miller 4584)

HAM RADIO

Fig. 2-40

This oscillator is designed for overtone crystals in the 20–100 MHz range operating in the third and fifth mode. Operating frequency is determined by the tuned circuit.

IC-COMPATIBLE CRYSTAL OSCILLATOR

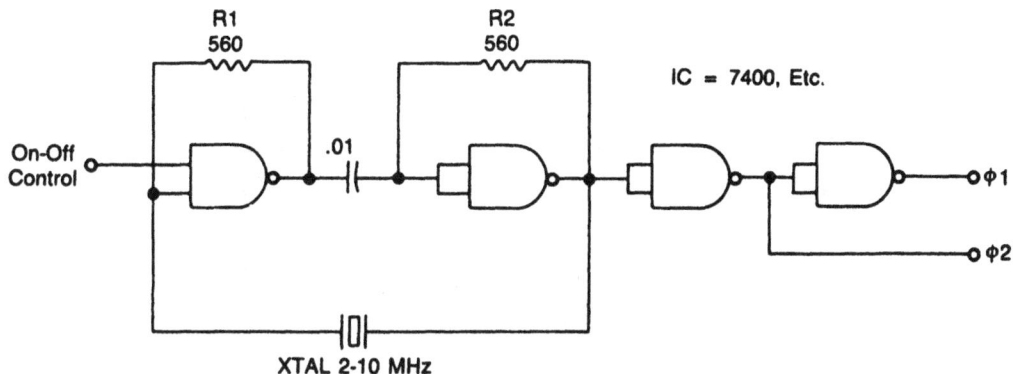

WILLIAM SHEETS

Fig. 2-41

Resistors R1 and R2 temperature-stabilize the NAND gates; they also ensure that the gates are in a linear region for starting. Capacitor C1 is a dc block; it must have less than $1/10$-Ω impedance at the operating frequency. The crystal runs in a series-resonant mode. Its series resistance must be low; AT-cut crystals for the 1- to 10-MHz range work well. The output waveshape has nearly a 50% duty cycle, with chip-limited rise times. The circuit starts well from 0° to 70°C.

CRYSTAL OSCILLATOR PROVIDES LOW NOISE

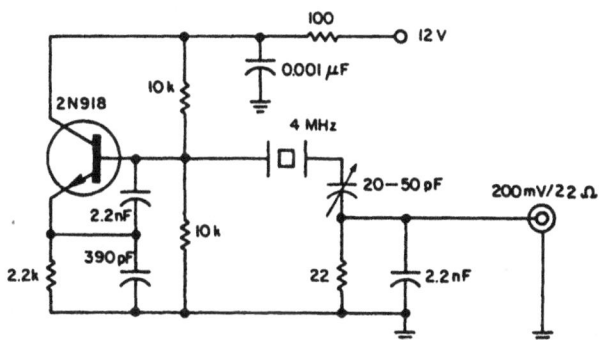

ELECTRONIC DESIGN

Fig. 2-42

The oscillator delivers an output of high spectral purity without any substantial sacrifice of the usual stability of a crystal oscillator. The crystal in addition to determining the oscillator's frequency, is used also as a low-pass filter for the unwanted harmonics and as a bandpass filter for the sideband noise. The noise bandwidth is limited to less than 100 Hz. All higher harmonics are substantially suppressed—60 dB down for the third harmonic of the 4-MHz fundamental oscillator frequency.

DISCRETE SEQUENCE OSCILLATOR

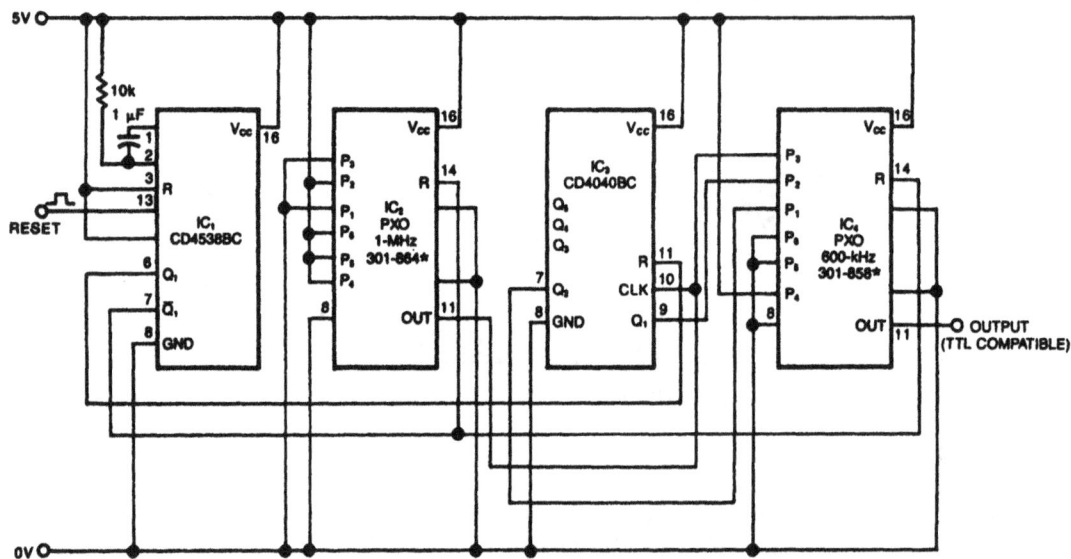

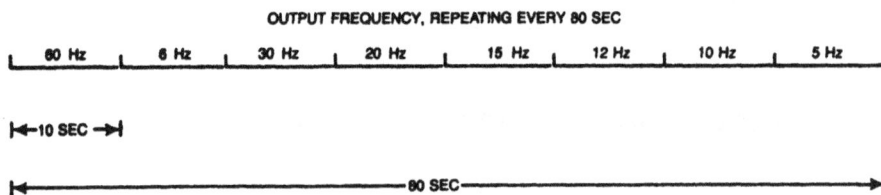

OUTPUT FREQUENCY, REPEATING EVERY 80 SEC

| 60 Hz | 6 Hz | 30 Hz | 20 Hz | 15 Hz | 12 Hz | 10 Hz | 5 Hz |

|◄—10 SEC —►|

|◄————————————————— 80 SEC —————————————————►|

*INTERFACE QUARTZ DEVICES LTD, CREWKERNE, SOMERSET, UK

EDN

Fig. 2-43

The swept-frequency oscillator offers an inexpensive source of discrete frequencies for use in testing digital circuits. In this configuration, the circuit generates an 80-second sequence of eight frequencies, dwelling for 10 seconds on each frequency. You can change the dwell time or the number of frequencies. Frequencies can range from 0.005 Hz to 1 MHz.

The programmable crystal oscillators, PXOs, IC2 and IC4 can each generate 57 frequencies in response to an 8-bit external code. IC2 contains a 1-MHz crystal and produces a 0.05-Hz output. IC4 contains a 600-kHz crystal; its output changes in response to the combined outputs of the 12-stage binary counter IC3 (Q1 and Q2) and the PXO IC2.

To generate more frequencies, you can use one or more of IC3's outputs, (Q3, Q4, Q5) to drive one or more of IC4's inputs (P4, P5, P6). Similarly, you can rewire IC2 or drive it with other logic to control the duration of each frequency. IC1, a monostable multivibrator, provides a system reset. It initiates the sequence shown, beginning at 60 Hz, in response to a positive pulse.

1-MHz PIERCE OSCILLATORS

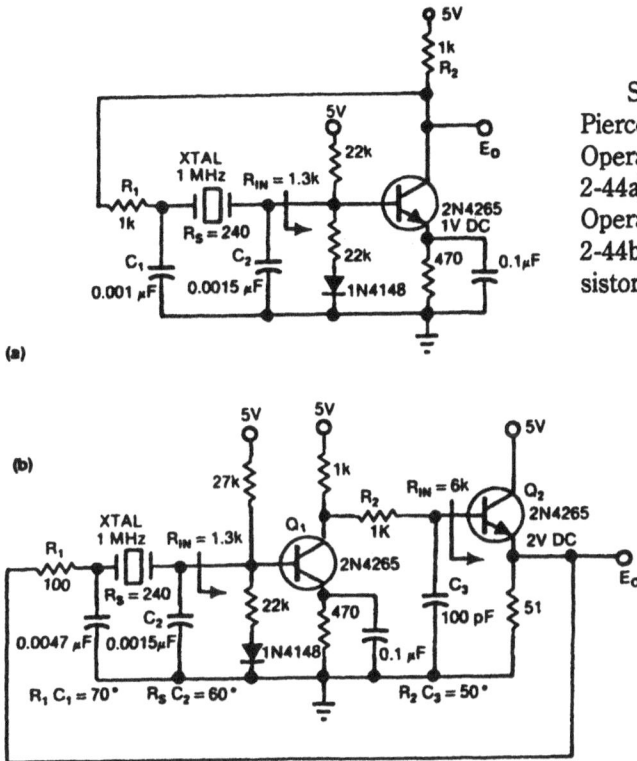

Simple network design is a key feature of the Pierce circuit, as these 1-MHz oscillators illustrate. Operating the crystal slightly above resonance (Fig. 2-44a) requires only one high-gain transistor stage. Operating it exactly at series resonance (Fig. 2-44b) requires an extra RC phase lag and two transistors which can have lower gain.

(a)

(b)

Fig. 2-44

SIMPLE CMOS CRYSTAL OSCILLATOR

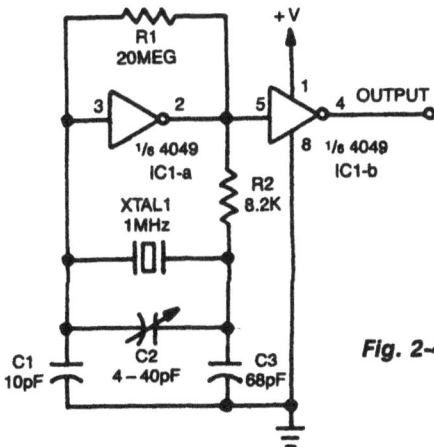

The circuit is an inverter set up as a linear amplifier. Adding the crystal and capacitors to the feedback path, we turn the amplifier into an oscillator and force it to oscillate at, or least very near, the crystal's resonant frequency. Trimmer capacitor C2 adjusts the actual operating frequency of the circuit. The crystal should be a parallel-resonant type; maximum frequency will depend partly on supply voltage, but it should be possible to go to at least 1 MHz.

Fig. 2-45

Reprinted with permission from Radio-Electronics Magazine February 1987. Copyright Gernsback Publications, Inc., 1987.

CRYSTAL TIMEBASE

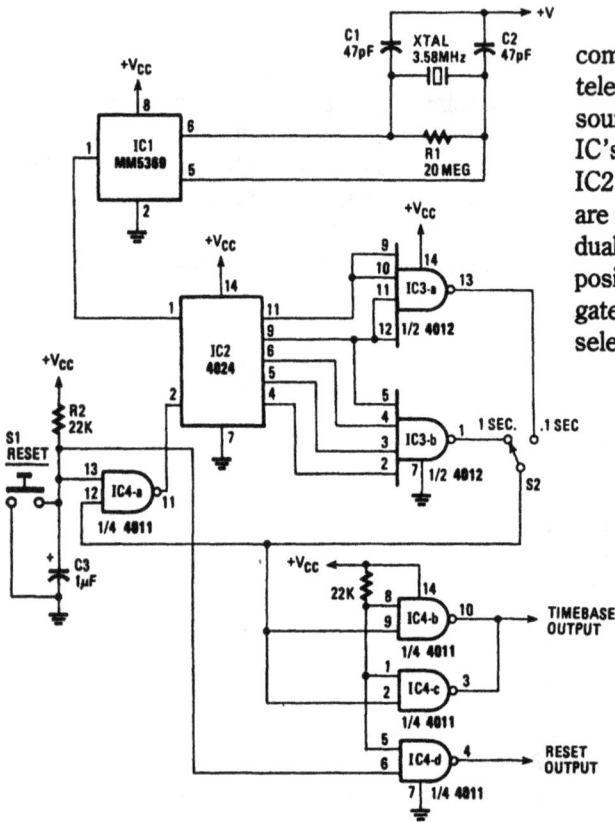

An on-board oscillator and a 17-stage divider compose IC1. By connecting a standard 3.58-MHz, television color-burst crystal as shown, an accurate source of 60-Hz squarewaves is generated at the IC's output, pin 1. Those pulses are then fed to IC2, a 4024 seven-stage ripple counter. Its outputs are connected to different gates in IC3, which is a dual four-input NAND gate. Depending on which position pulse-select switch S2 is on, one of those gates will provide an output/reset pulse of the selected width.

Fig. 2-46

LOW-FREQUENCY PIERCE OSCILLATOR

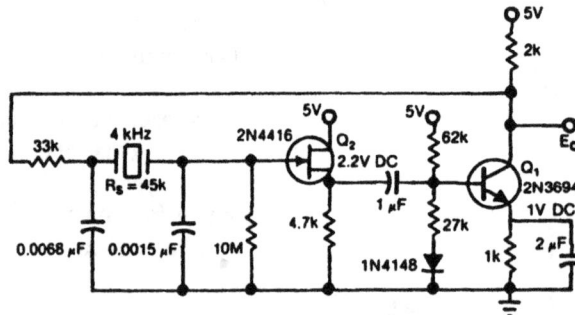

EDN

Fig. 2-47

The Pierce circuit oscillates at 4 kHz. At low frequencies, the crystal's internal series resistance R_S is quite high (45 K at 4 kHz). Therefore, an FET-based source follower is included to prevent Q1 from loading the crystal output.

1-MHz FET CRYSTAL OSCILLATOR

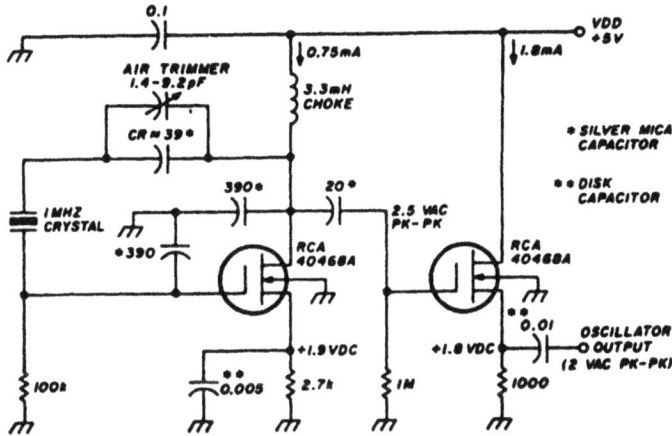

HAM RADIO

Fig. 2-48

This stable oscillator circuit exhibits less than 1-Hz frequency change over a V_{DD} range of $3-9$ volts. Stability is attributed to the use of MOSFET devices and the use of stable capacitors.

PIERCE CRYSTAL OSCILLATOR

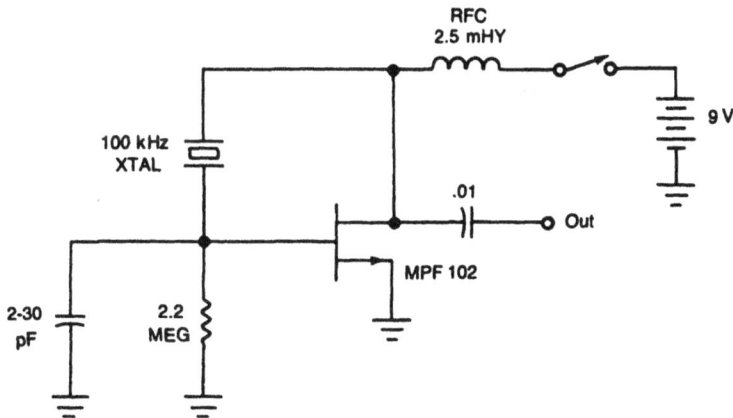

WILLIAM SHEETS

Fig. 2-49

The JFET Pierce oscillator is stable and simple. It can be the clock of a microprocessor, a digital time-piece, or a calculator. With a probe at the output, it can be used as a precise injection oscillator for trouble-shooting. Attach a small length of wire at the output and this circuit becomes a micropower transmitter.

OVERTONE OSCILLATOR WITH CRYSTAL SWITCHING

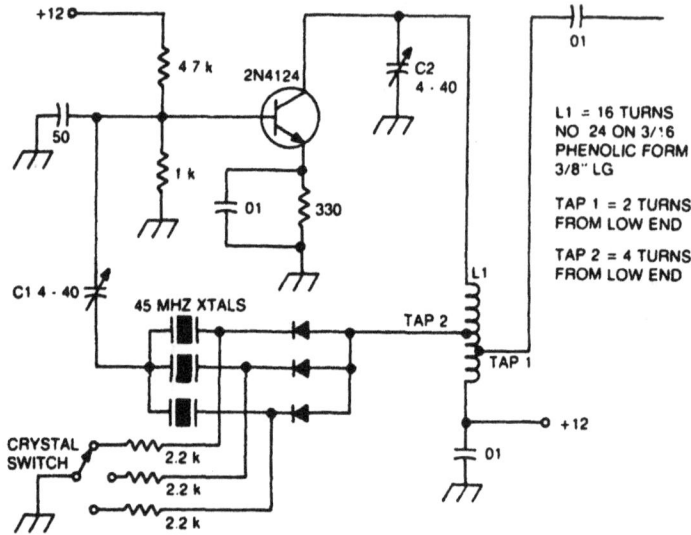

L1 = 16 TURNS
NO 24 ON 3/16
PHENOLIC FORM
3/8" LG

TAP 1 = 2 TURNS
FROM LOW END

TAP 2 = 4 TURNS
FROM LOW END

LINEAR TECHNOLOGY CORP.

Fig. 2-50

The large inductive phase shift of L1 is compensated for by C1. Overtone crystals have very narrow bandwidth; therefore, the trimmer has a smaller effect than for fundamental-mode operation.

CRYSTAL OSCILLATOR

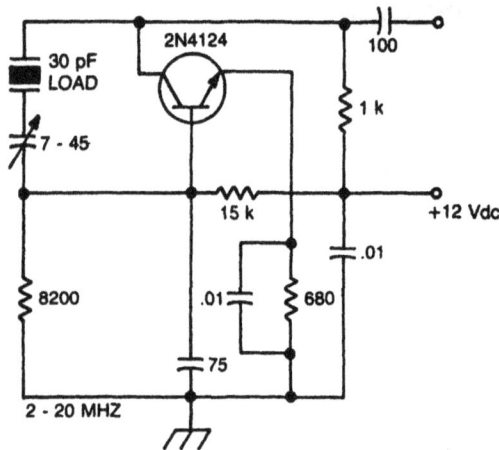

The crystal is in a feedback circuit from collector to base. A trimmer capacitor in series shifts the point on the reactance curve where the crystal operates, thus providing a frequency trim. The capacitor has a negative reactance so that the crystal is shifted to operate in the positive reactance region.

EDN

Fig. 2-51

FIFTH-OVERTONE OSCILLATOR

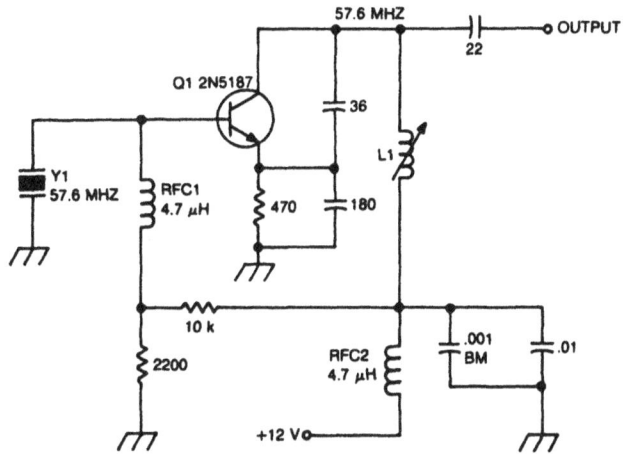

Fig. 2-52

This circuit isolates the crystal from the dc base supply with an rf choke for better starting characteristics.

CRYSTAL-CONTROLLED BUTLER OSCILLATOR

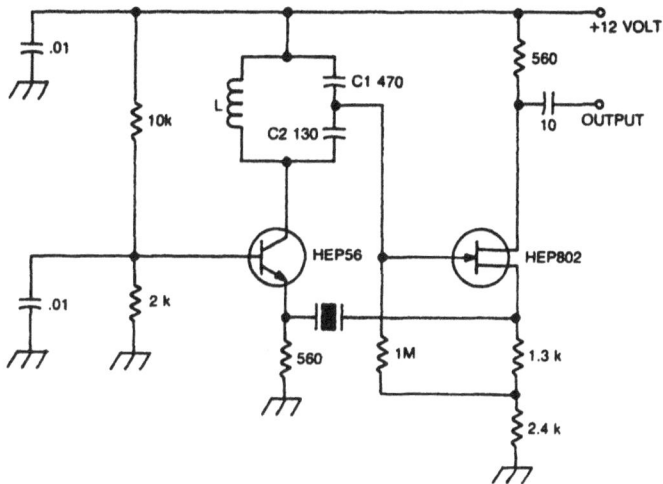

Fig. 2-53

A typical Butler oscillator (20 – 100 MHz) uses an FET in the second stage; the circuit is not reliable with two bipolars. Sometimes two FETs are used. Frequency is determined by LC values.

SCHMITT TRIGGER CRYSTAL OSCILLATOR

SCHMITT TRIGGER OSCILLATOR UP TO 10 MHZ

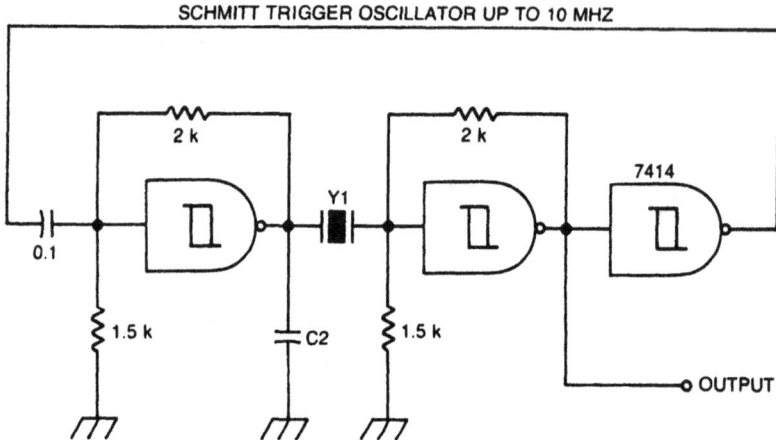

Fig. 2-54   NOTE: C2 = 1/1 × 10⁻⁴ (f IS IN HZ) - PREVENTS SPURIOUS FREQUENCY   **HAM RADIO**

A Schmitt trigger provides good squaring of the output, and sometimes eliminates the need for an extra output stage.

OVERTONE OSCILLATOR (50 TO 150 MHz)

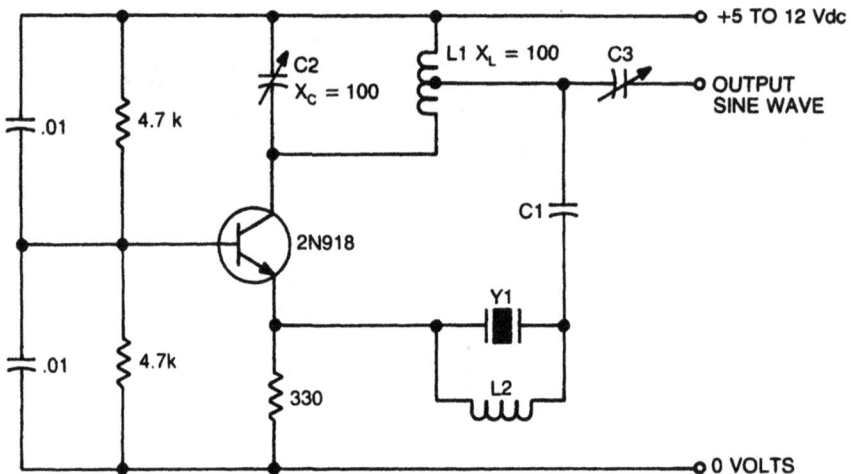

NOTES:
1. Y1 IS AT CUT OVERTONE CRYSTAL.
2. TUNE L1 AND C2 TO OPERATING FREQUENCY.
3. L2 AND SHUNT CAPACITANCE, CO. OF CRYSTAL (APPROXIMATELY 6pF) SHOULD RESONATE TO OSCILLATOR OUTPUT FREQUENCY (L2 = .5 μH AT 90 MHZ). THIS IS NECESSARY TO TUNE OUT EFFECT OF CO.
4. C3 IS VARIED TO MATCH OUTPUT.

HAM RADIO   **Fig. 2-55**

PRECISION CLOCK GENERATOR

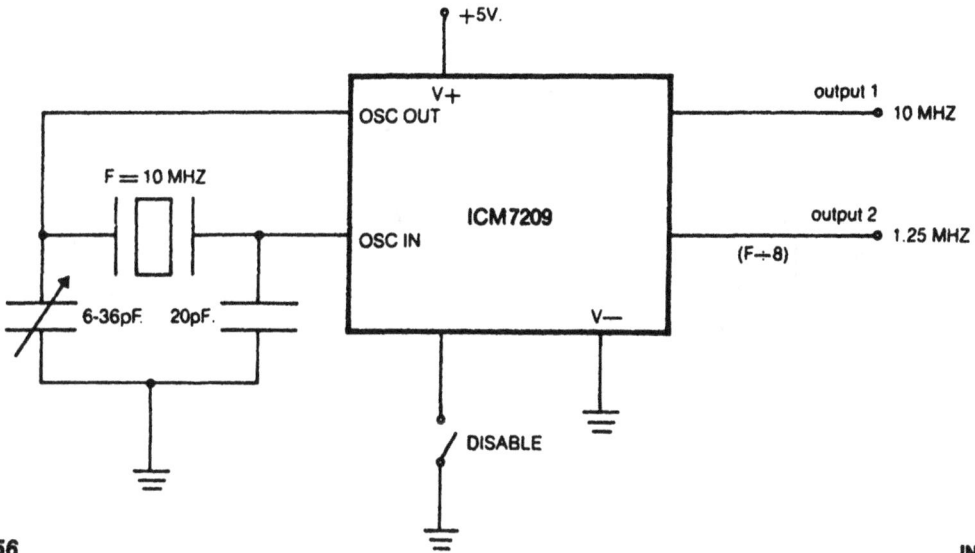

Fig. 2-56

INTERSIL

The CMOS IC directly drives 5 TTL loads from either of 2 buffered outputs. The device operates to 10 MHz and is bipolar, MOS, and CMOS compatible.

MILLER OSCILLATOR (CRYSTAL CONTROLLED)

The drain of the JFET Miller oscillator is tuned to the resonant frequency of the crystal by an LC tank circuit.

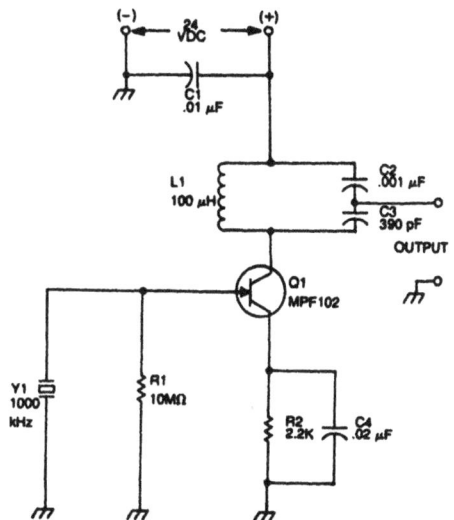

TAB BOOKS

Fig. 2-57

BUTLER EMITTER-FOLLOWER OSCILLATOR (BASIC CIRCUIT)

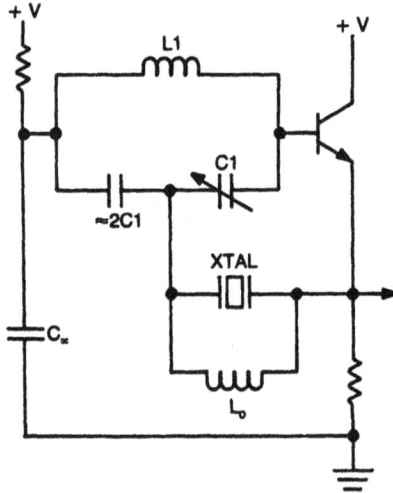

This circuit operates at or near series resonance. It is a good circuit design with no parasitics. It is easy to tune with good frequency stability.

JOHN WILEY & SONS   *Fig. 2-58*

BUTLER COMMON-BASE OSCILLATOR (BASIC CIRCUIT)

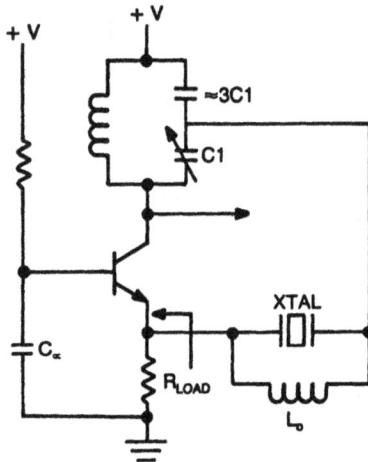

This circuit operates at or near series resonance. It has fair to poor circuit design with parasitics, touch to tune, and fair frequency stability.

JOHN WILEY & SONS   *Fig. 2-59*

CMOS OSCILLATOR (1 TO 4 MHz)

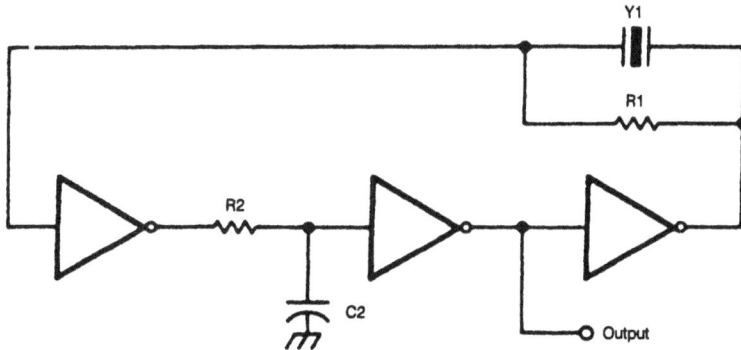

Notes:
1. IM < R1 < 5M
2. Select R2 and C2 to prevent spurious frequencies
3. ICs are 74C04 or equivalent

HAM RADIO

Fig. 2-60

PIERCE HARMONIC OSCILLATOR (100 MHz)

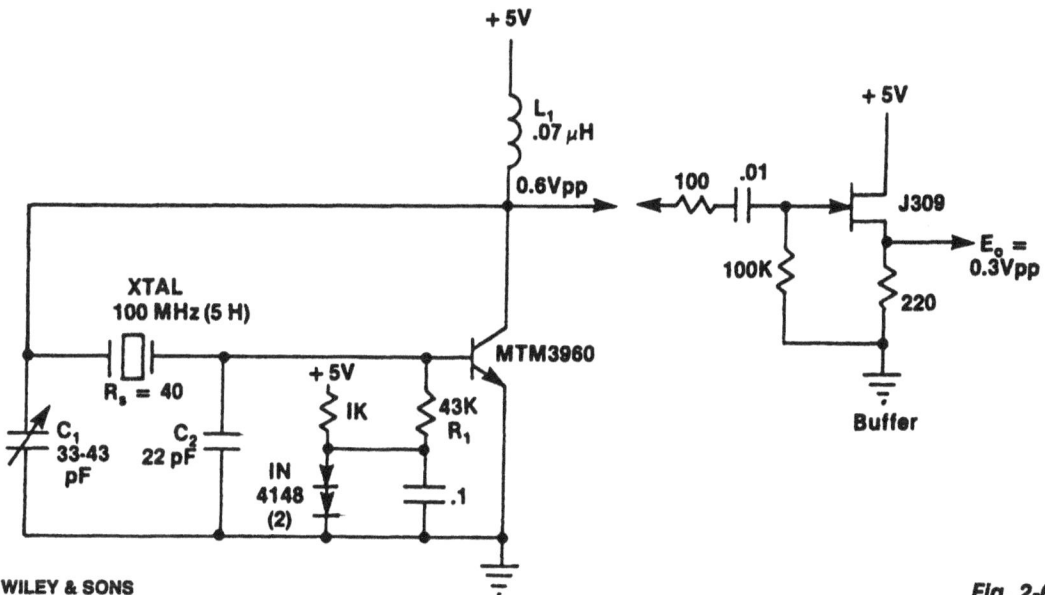

JOHN WILEY & SONS

Fig. 2-61

The output resistance of the transistor's collector, together with the effective value of C1, provides an RC phase lag of 30 – 50°. The crystal normally oscillates slightly above series resonance, where it is both resistive and inductive. Above series resonance, the crystal's internal impedance (resistive and inductive) together with C2 provides an RLC phase lag of 130 – 150°. The transistor inverts the signal, providing a total of 360° of phase shift around the loop. Inductor L1 is selected to resonate with C1 at a frequency between the crystal's desired harmonic and its next lower odd harmonic. Inductor L1 offsets part of the negative reactance of C1 at the oscillation frequency.

INTERNATIONAL CRYSTAL OF-1 HI OSCILLATOR

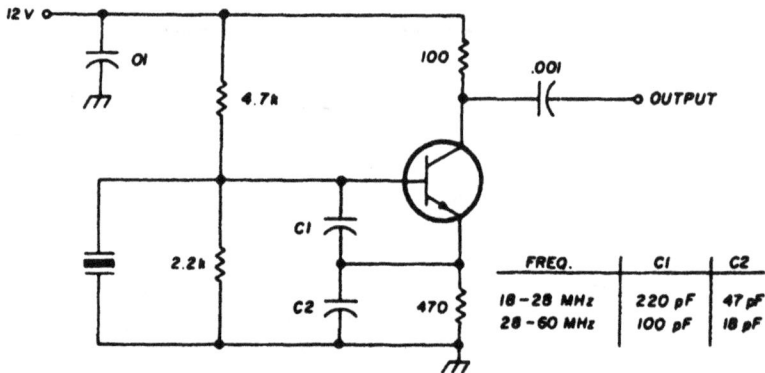

FREQ.	CI	C2
18–28 MHz	220 pF	47 pF
28–60 MHz	100 pF	18 pF

HAM RADIO

Fig. 2-62

International Crystal OF-1 HI oscillator circuit for third-overtone crystals. The circuit does not require inductors.

STANDARD CRYSTAL OSCILLATOR FOR 1 MHz

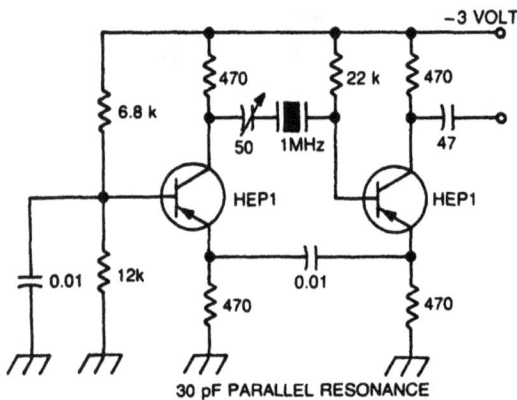

30 pF PARALLEL RESONANCE

HAM RADIO

Fig. 2-63

TTL-COMPATIBLE CRYSTAL OSCILLATOR

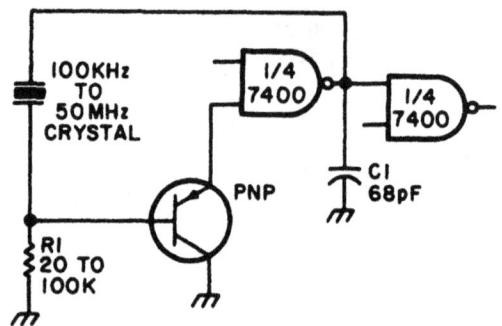

73 AMATEUR RADIO

Fig. 2-64

Adjust R1 for about 2 volts at the output of the first gate. Adjust C1 for best output.

THIRD-OVERTONE CRYSTAL OSCILLATOR

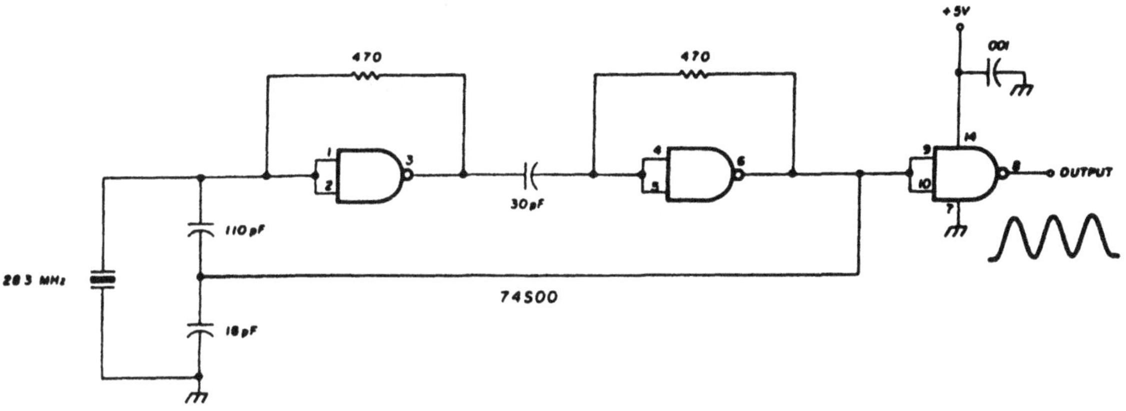

HAM RADIO

Fig. 2-65

This circuit uses a 74S00 Schottky TTL gate; no inductors are required.

CRYSTAL CHECKER

This circuit is a simple Pierce oscillator with an LED go/no go display. The checker works best with crystals having fundamental frequencies in the 7 to 8 MHz range.

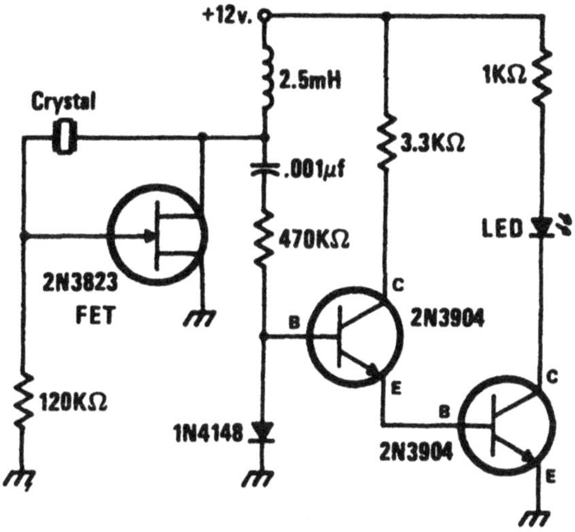

MODERN ELECTRONICS

Fig. 2-66

CRYSTAL OSCILLATOR/DOUBLER

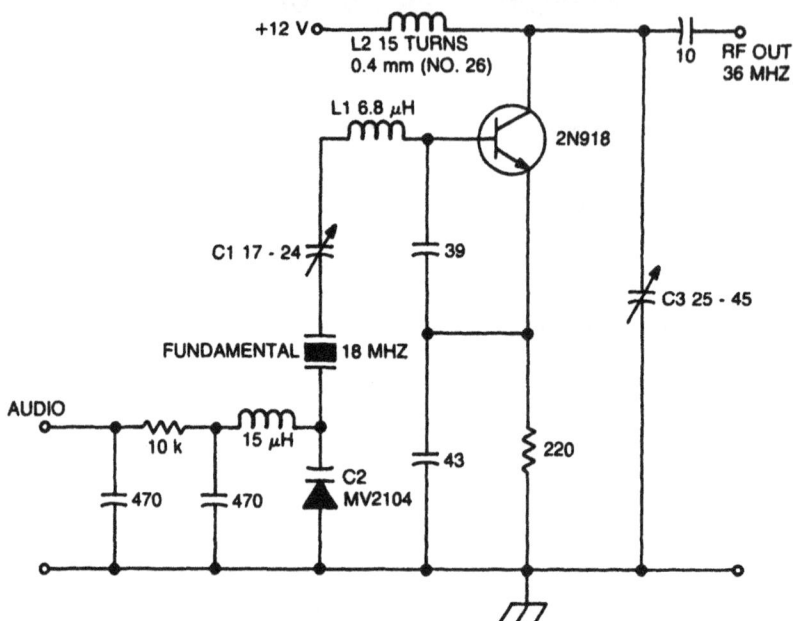

Fig. 2-67

The crystal operates into a complex load at series resonance. L1, C1, and C2 balance the crystal at zero reactance. Capacitor C1 fine-tunes the center frequency. Tank circuit L2/C3 doubles the output frequency. The circuit operates as an FM oscillator-doubler.

LOW-FREQUENCY CRYSTAL OSCILLATOR

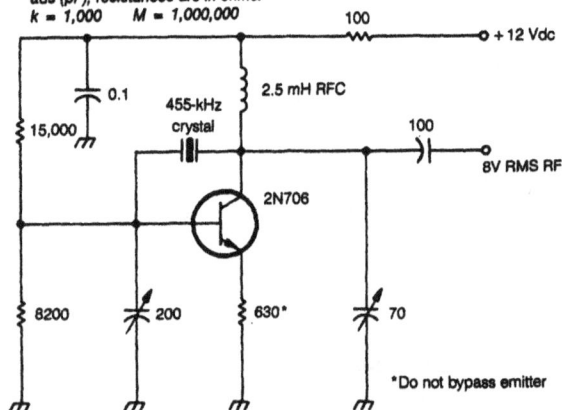

This crystal-oscillator circuit uses a 455-kHz crystal.

HAM RADIO

Fig. 2-68

VARACTOR-TUNED 10-MHz CERAMIC RESONATOR OSCILLATOR

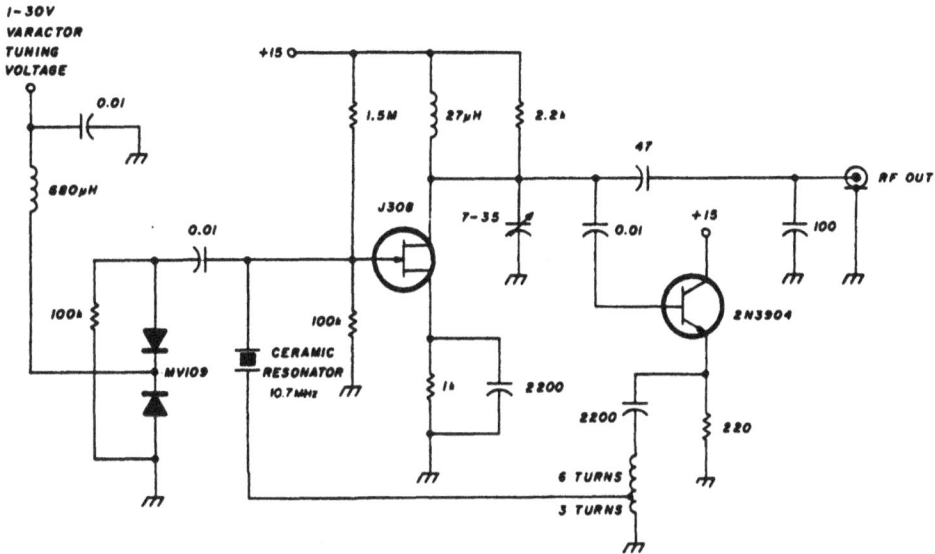

HAM RADIO

Fig. 2-69

The FET input amplifier has fixed bias with source feedback. This provides a very high input impedance with very low capacitance. The FET amplifier drives an emitter follower, which, in spite of the fact that it has a low output impedance, feeds a transformer with a 3:1 turns ratio for a nine-fold impedance reduction. The result is an impedance at the ceramic resonator of a few ohms maximum. The varactor-tuned ceramic resonator oscillator has a significant frequency-temperature coefficient. The tuning range of the VCO is approximately 232 kHz, with a temperature coefficient of 350 Hz per degree centigrade. When using this circuit as a VCO, the entire 232-kHz range cannot be used because some of the tuning range must be sacrificed for the temperature dependence. If the required tuning range was 200 kHz (leaving 32 kHz for temperature variation), the resulting temperature variation would be more than 90°C.

10-MHz CRYSTAL OSCILLATOR

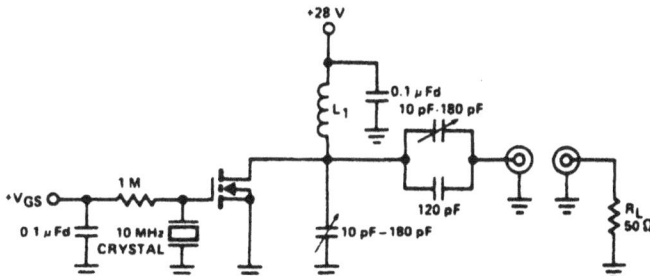

SILICONIX

Fig. 2-70

This crystal oscillator is a FET equivalent of a vacuum tube tuned to plate-tuned grid crystal oscillator. Feedback is via the drain to gate capacitance.

PIERCE HARMONIC OSCILLATOR (BASIC CIRCUIT)

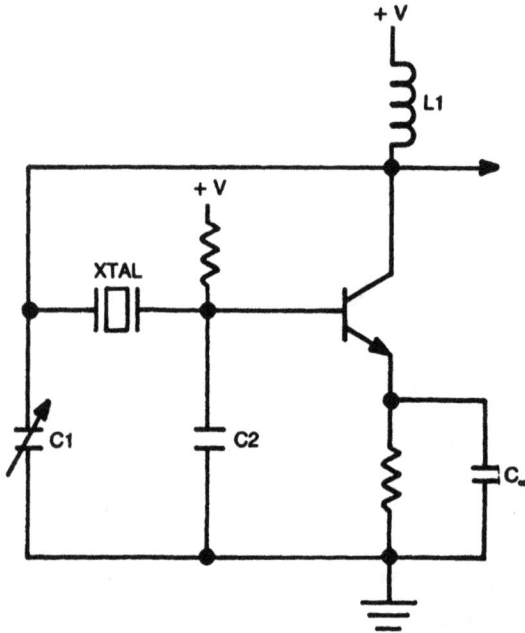

This circuit operates 10–40 ppm above series resonance. It is a good circuit design with good to very good frequency stability.

JOHN WILEY & SONS   *Fig. 2-71*

TUBE-TYPE CRYSTAL OSCILLATOR

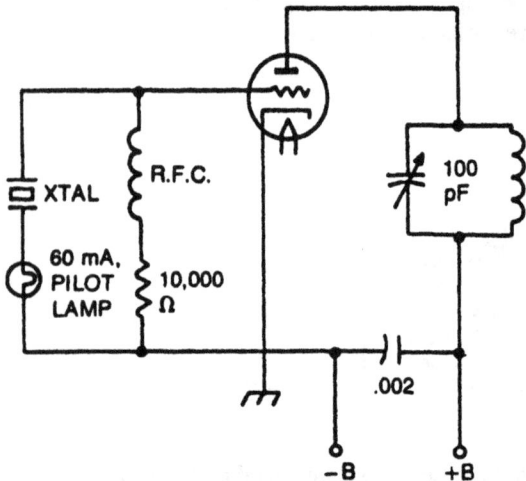

The pilot lamp limits current to prevent damage to the crystal.

TAB BOOKS   *Fig. 2-72*

CRYSTAL-CONTROLLED SINE-WAVE OSCILLATOR

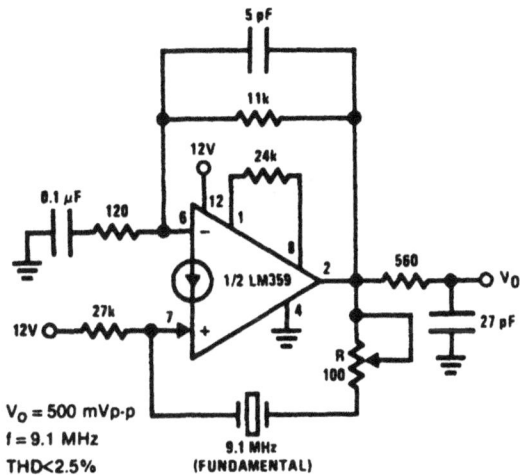

V_O = 500 mVp-p
f = 9.1 MHz
THD<2.5%

NATIONAL SEMICONDUCTOR

Fig. 2-73

STABLE LOW-FREQUENCY CRYSTAL OSCILLATOR

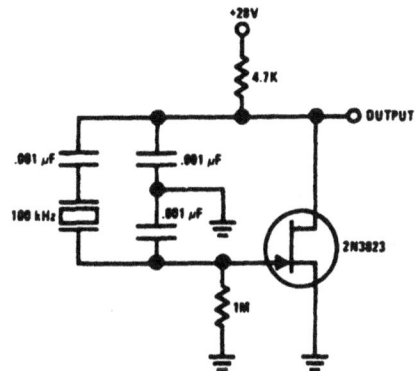

NATIONAL SEMICONDUCTOR   *Fig. 2-75*

This Colpitts-crystal oscillator is ideal for low-frequency crystal oscillator circuits. Excellent stability is assured because the 2N3823 JFET circuit loading does not vary with temperature.

CRYSTAL OSCILLATOR

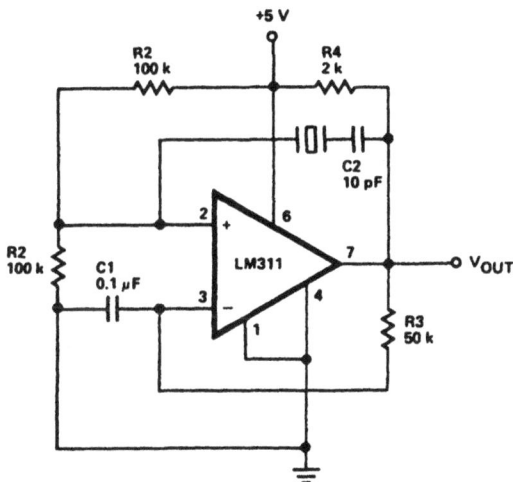

TELEDYNE

Fig. 2-74

JFET PIERCE CRYSTAL OSCILLATOR

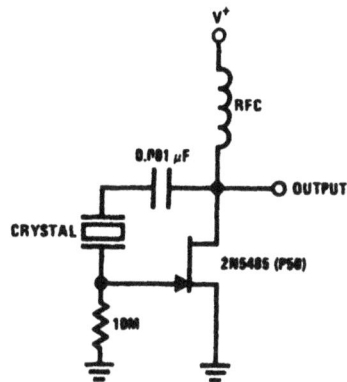

NATIONAL SEMICONDUCTOR   *Fig. 2-76*

The JFET Pierce crystal oscillator allows a wide frequency range of crystals to be used without circuit modification. Because the JFET gate does not load the crystal, good Q is maintained, thus ensuring good frequency stability.

CRYSTAL OSCILLATOR

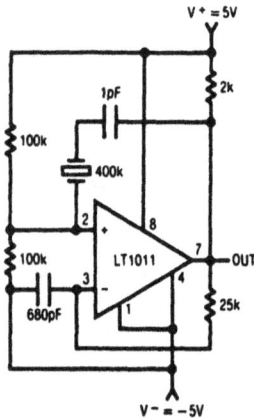

This circuit uses an LT1011 comparator biased in its linear mode and a crystal to establish its resonant frequency. This circuit can achieve a few hundred kHz, and it has temperature independent clock frequency with nearly 50% duty cycle.

Fig. 2-77   LINEAR TECHNOLOGY CORP.

CRYSTAL-CONTROLLED SIGNAL SOURCE

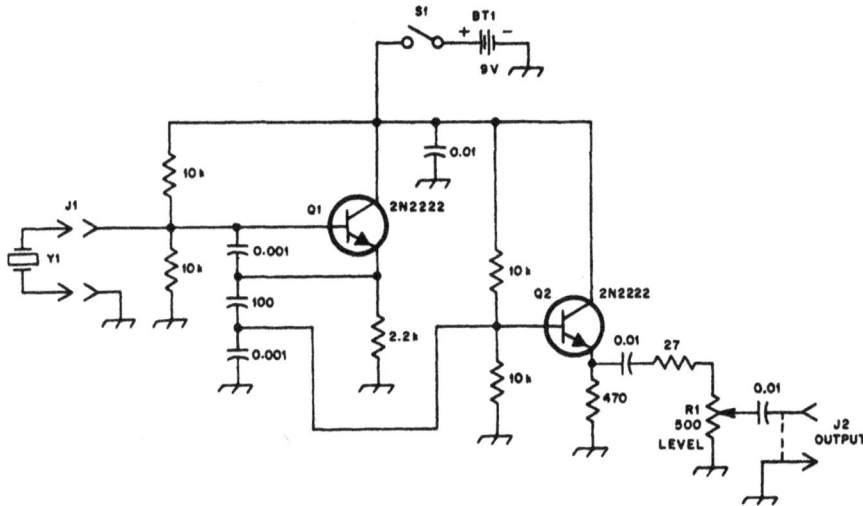

QST

Fig. 2-78

This general-purpose signal source serves very well in signal-tracing applications. The output level is variable to more than 1 Vrms into a 50-Ω load. Almost any crystal in the 1- to 15-MHz range can be used. Q1 forms a Colpitts oscillator with the output taken from the emitter. A capacitive voltage divider (across the 2.2-kΩ emitter resistor) reduces the voltage applied to the buffer amplifier, Q2. The buffer and emitter follower, provides the low-input impedance necessary to drive 50-Ω loads.

HIGH-FREQUENCY SIGNAL GENERATOR

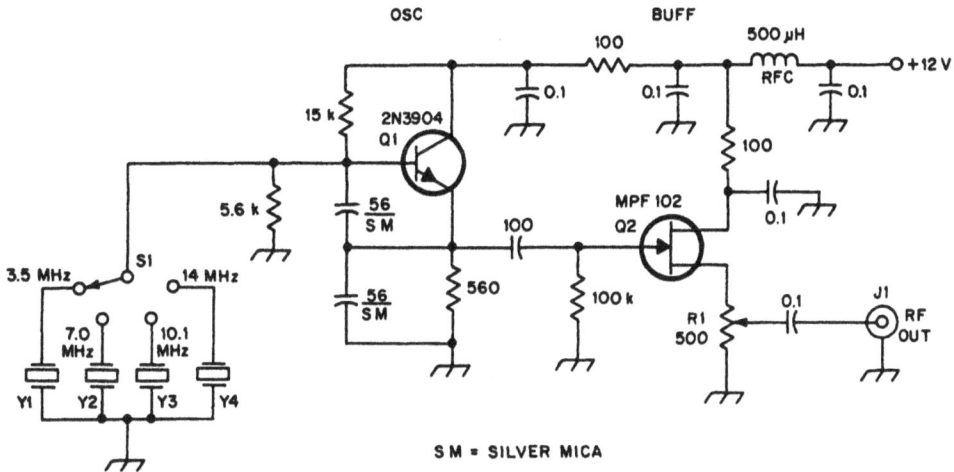

S M = SILVER MICA

QST

Fig. 2-79

A switched-crystal Colpitts oscillator is used at Q1 to provide four tuning ranges from 1.7 to 3.1 MHz, 3.0 to 5.6 MHz, 5.0 to 12 MHz and 11.5 to 31 MHz. A Zener diode (D2) is used at Q1 to lower the operating voltage of the oscillator. A small-value capacitor is used at C5 to ensure light coupling to the tuned circuit. Q2 is a source-follower buffer stage. It helps to isolate the oscillator from the generator-output load. The source of Q2 is broadly tuned by means of RFC1. Energy from Q2 is routed to a fed-back, broadband class-A amplifier. A 2-dB attenuator is used at the output of T1 to provide a 50-Ω termination for Q3 and to set the generator-output impedance at 50 Ω. C16, C17 and RFC2 form a brute-force RF decoupling network to keep the generator energy from radiating outside the box on the 12-V supply.

CRYSTAL-STABILIZED IC TIMER CAN
PROVIDE SUBHARMONIC FREQUENCIES

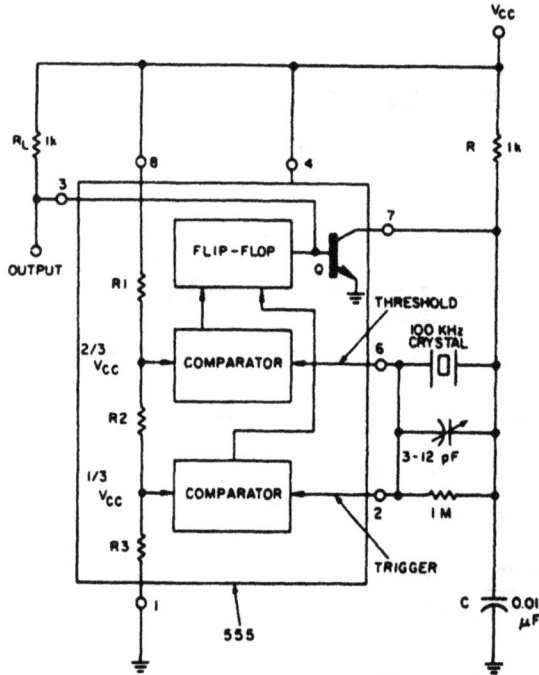

The trimmer across the crystal can finely tune the circuit's oscillating frequency.

ELECTRONIC DESIGN

Fig. 2-80

TTL OSCILLATOR FOR 1 TO 10 MHz

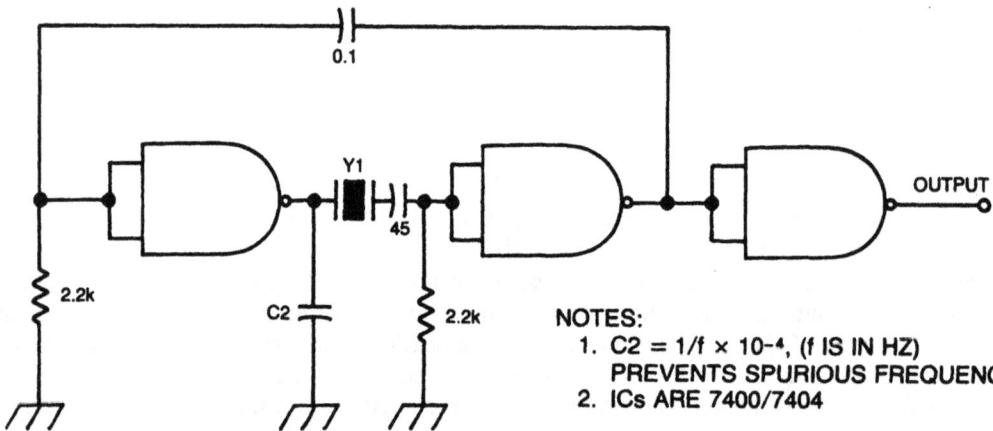

NOTES:
1. $C2 = 1/f \times 10^{-4}$, (f IS IN HZ) PREVENTS SPURIOUS FREQUENCY
2. ICs ARE 7400/7404

HAM RADIO

Fig. 2-81

50-MHz VHF CRYSTAL OSCILLATOR

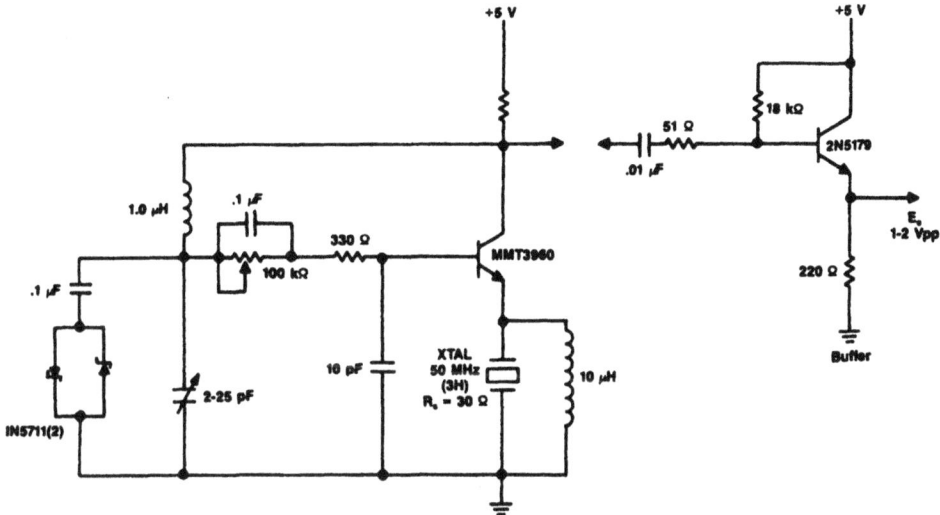

R. Matthys, RF Design, March 1987, p. 31.

Fig. 2-82

Figure 2-82 shows a 50-MHz oscillator operating on a third harmonic. The collector's load resistor R1 has been increased because the quartz crystal's internal series resistance R_s increases with frequency in the VHF range. The crystal's internal series resistance, R_s, is 30 Ω, and the transistor's minimum current gain H_{FE} is 100. Using the same technique as for the 20-MHz oscillator, the external series $R_L C_L$ equivalent load seen by the 50-MHz crystal is 5.6 Ω (R_L) and 1000 pF (C_L).

3

Function Generators

The sources of the following circuits are contained in the Sources section, which begins on page 173. The figure number contained in the box of each circuit correlates to the source entry in the Sources section.

Sine/Cosine Generator (0.1 to 10 kHz)

Quad Op Amp Synchronized Function Generator

Oscillator or Amplifier with Wide Frequency Range

Linear Triangle-/Square-Wave VCO

Circuit for Multiplying Pulse Widths

Programmable Voltage-Controlled Frequency
   Synthesizer

Emitter-Coupled RC Oscillator

Voltage-Controlled High-Speed One Shot

Ramp Generator with Variable Reset Level

555 Astable with Low Duty Cycle

Monostable Using Video Amplifier and Comparator

UJT Monostable Circuit Insensitive to Change in
   Bias Voltage

Linear Ramp Generator

Function Generator

Astable Multivibrator

Waveform Generator

Last-Cycle Completing Gated Oscillator

Precise Triangle-/Square-Wave Generator

Wide-Range Triangle-/Square-Wave Generator

Waveform Generator

Single-Supply Function Generator

Precise Wave Generator

Fixed-Frequency Variable Duty-Cycle Oscillator

Wide-Range Variable Oscillator

Single-Control Four-Decade Variable Oscillator

Simple Triangle Square-Wave Oscillator

PRECISION ONE SHOT

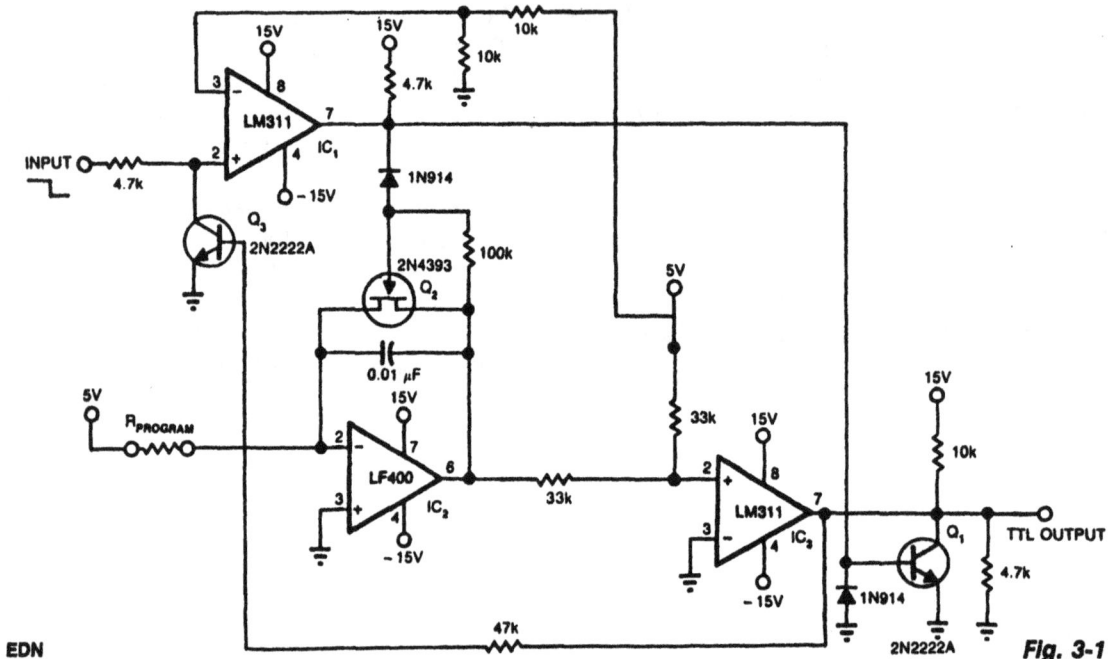

Fig. 3-1

EDN

If you need a wide-range, resistor-programmable monostable multivibrator, you can program the circuit for pulse widths from 1 μs to 10 s ($-.10^7$:1 range). A high-to-low transition at the input causes IC1's output to switch low, thereby turning off Q1 and Q2. With the latter transistor turned off, IC3's output increases and the output of IC2 begins to ramp toward the negative supply level at a rate determined by the 0.01-μF capacitor and the programming resistor. When IC2's output voltage reaches -5 V, IC3's output switches low. If you anticipate input pulses shorter than the desired output pulses, Q3 is necessary. This transistor keeps IC1s input low while an output pulse is present, thereby preventing inadvertent resetting of the one shot.

LINEAR TRIANGLE-WAVE TIMER

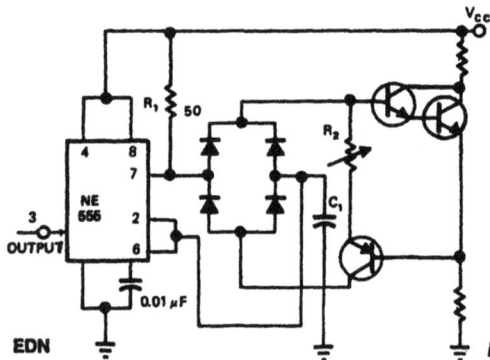

Fig. 3-2

EDN

Using one current source for the charge and discharge path in this circuit ensures identical rise and fall times at the capacitor terminal. A Darlington pair ensures identical biasing of the IC during the charge and discharge cycles. The period of the triangle wave is: $T \approx 0.46 V C_1/R_2$. V_{CC} must be at least 8 V to maintain linearity. At the output at pin 3 of the IC timer, a 50% duty-cycle square wave, frequency tunable by R2 alone, appears.

FOUR-OUTPUT WAVEFORM GENERATOR

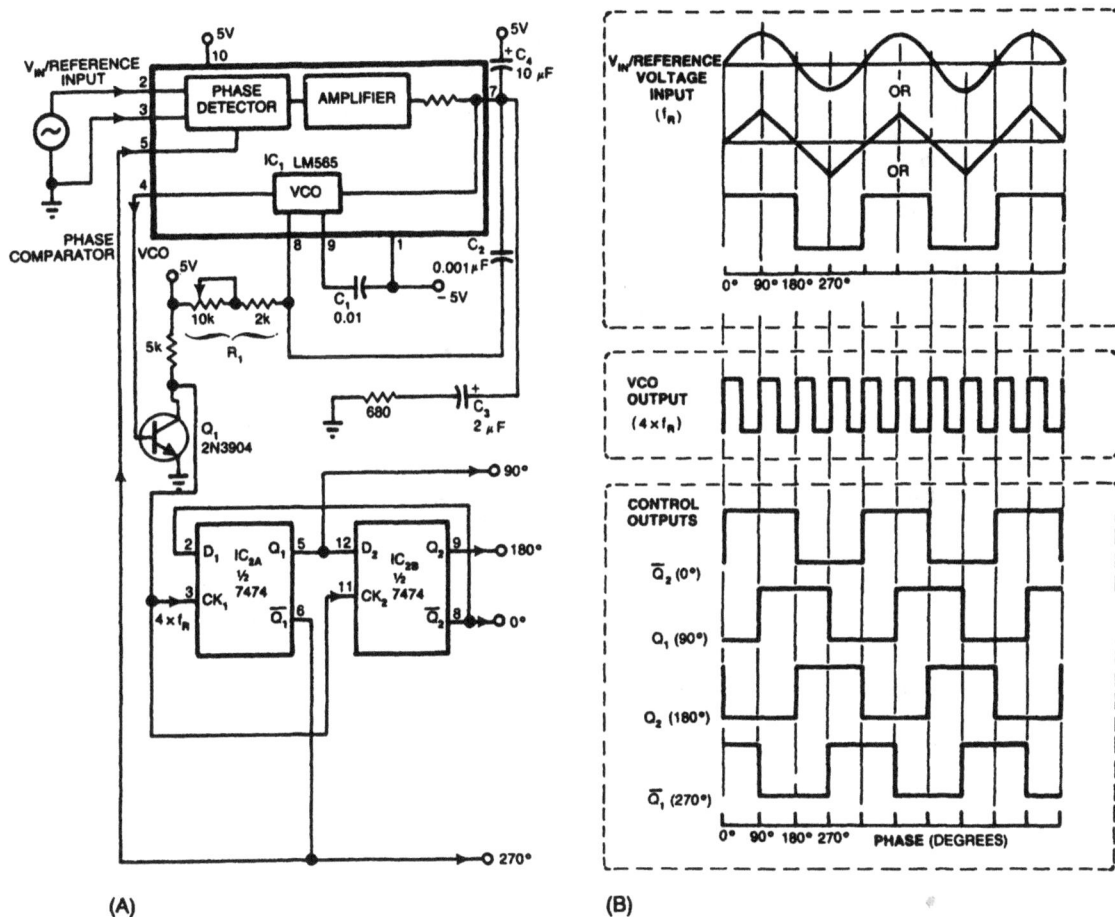

(A)   (B)

EDN

Fig. 3-3

Many applications require control signals that have phase shifts with reference to an input signal. Circuit accepts a sine, square, or triangular wave as an input reference signal and produces square-wave outputs with 0°, 90°, 180°, and 270° phase shifts with respect to the input. Figure 3-3B shows the input and output waveforms. The circuit contains two ICs: an LM565 phase-locked loop and a 7474 dual-D positive edge-triggered flip-flop. R1 and C1 set the free-running frequency of the LM565's VCO. You should adjust R1 so that the frequency is approximately four times that of the input reference signal. The LM565 responds to input signals greater than 10 mV pk-pk; 3 V pk-pk is the chip's maximum allowable input level. Q1 matches the LM565's output to the flip-flops' inputs. The flip-flops' outputs provide the TTL-compatible square-wave signals with 0°, 90°, 180°, and 270° phase shift, with reference to the input signal.

FUNCTION GENERATOR

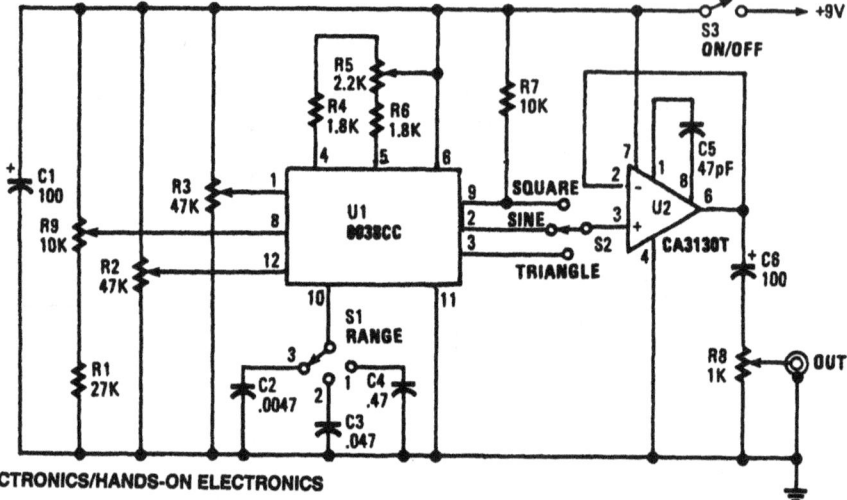

Fig. 3-4

This circuit can output sine, square, and triangular signals from 15 Hz to 25 kHz in three ranges. The circuit is built around an 8038 function generator that produces the triangular- and square-wave outputs directly from an oscillator. The triangular output is then processed to develop the sine wave. While that method doesn't provide a sufficiently low level of distortion to let you make distortion measurements on audio gear, the degree of purity is high enough for frequency-response tests and a lot of other audio analysis. Three switched capacitors, C2 to C4, set the circuits frequency range via switch S1. Variable resistor R9 and resistor R1 provide the voltage for controlling the charge and discharge rates of the timing capacitor selected. Resistors R4 to R6 control the charge and discharge currents. Resistor R5 can be adjusted to provide a 1.1 mark/space ratio.

CLASSIC OP-AMP ASTABLE MULTIVIBRATOR

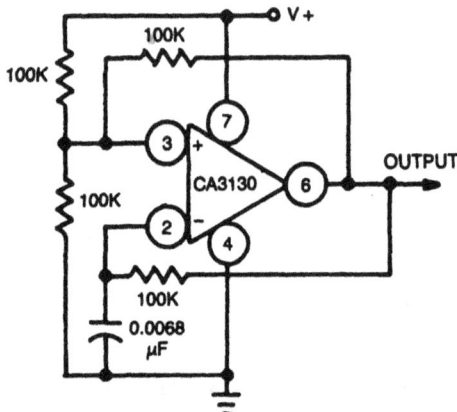

Uses CA3130 BiMOS op amp that operates at a frequency of 1 kHz. With rail-to-rail output swing, frequency is independent of supply voltage, device, and temperature. Only the temperature coefficient of R_T and C_T enters into circuit stability.

GE/RCA   ALL RESISTANCE VALUES ARE IN OHMS   *Fig. 3-5*

PROGRAMMABLE TRIANGLE-/SQUARE-WAVE GENERATOR

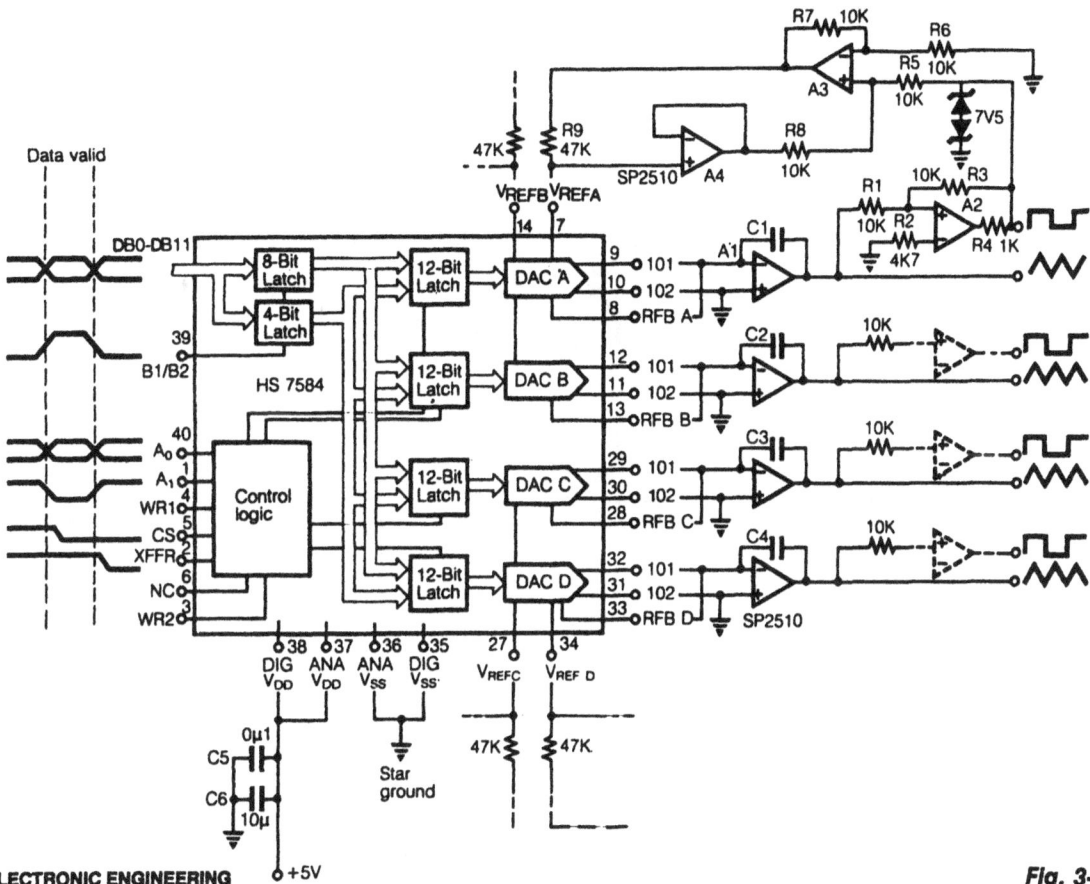

ELECTRONIC ENGINEERING

Fig. 3-6

The programmable multiple output generator provides the control signals for data converter ATE. Major performance criteria are simple interfaces to a number of microprocessor systems, low power consumption, stable output timing relationships combined with a minimum of board space. For schematic simplicity only, one output circuit is shown in full.

The monolithic HS7584 provides four current output DAC's with four quadrant multiplication, individual reference input and a feedback resistor. The digitally controlled integrator's frequency is determined by:

$$f = \frac{\text{Digital Input}}{4RC}$$

C is the value of C1 to C4 and R is the resistance of the DAC. With the four DACS on a single chip, the resistance matching is good, which results in stable timing relationships of the generator outputs. The output of the comparator A2 determines whether the constant current source provided by A3 and A4 is positive or negative.

NONINTEGER PROGRAMMABLE PULSE DIVIDER

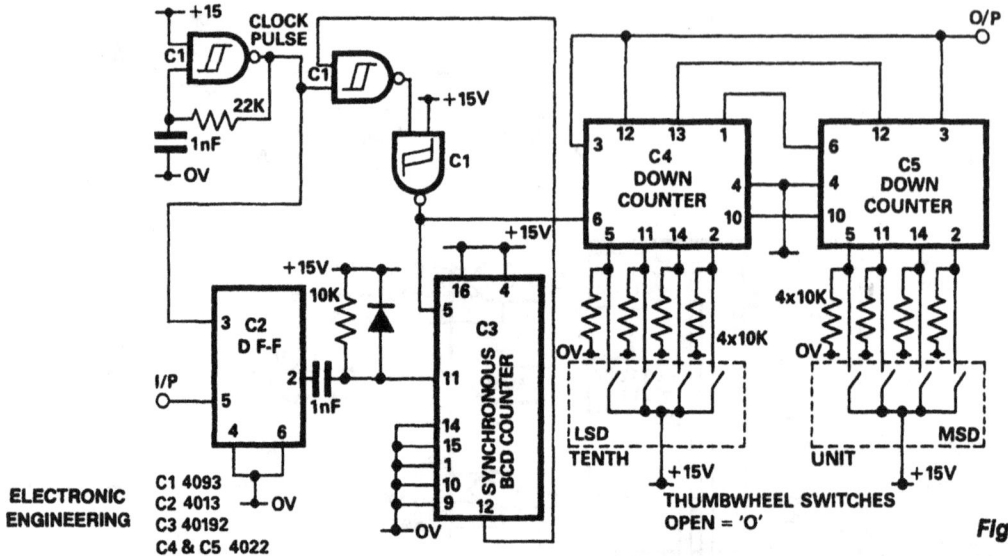

Fig. 3-7

The purpose of D-type flip-flop IC2 is to synchronize the input signal with the clock pulse. When the clock pulse changes from low to high and the input is high, IC2 output is high. Subsequently, IC3 resets to zero and starts counting up. Until the counter counts to ten, the counter is inhibited. Thus, the number of pulses of the output of IC3 is ten times input pulse. The designed frequency of the clock pulse must be ten times higher than the maximum frequency of the input. IC4 and IC5 are cascaded to form a two decade programmable down counter. Since the number of pulses appearing at the input of the down counter is ten times the input to the divider, the effective range of the divisor for this divider is 0.1 to 9.9.

XOR-GATE COMPLEMENTARY-SIGNALS GENERATOR

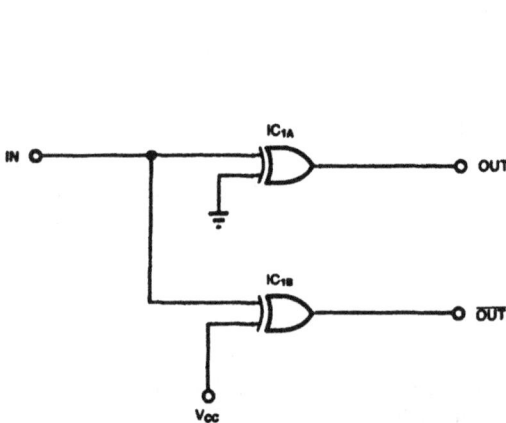

EDN   Fig. 3-8

Some applications, such as driving three-state buffers for data multiplexers or for biphase clocks in high-speed systems, require complementary signals having a small-time skew and nearly simultaneous transitions. Here, XOR gates function as both inverting and noninverting gates. For CMOS systems, practically any type of XOR gate will work. However, the advanced-CMOS logic (ACL) families have the greatest drive capability, the shortest gate delays, and the tightest manufacturing tolerances. For TTL systems, compatible CMOS types such as the ACT or S/AS86 families are preferable. Do not use low-power TTL, versions (LS or ALS), because they have large propagation delay differences when one XOR gate is inverting and the other is noninverting.

LOW-COST FSK GENERATOR

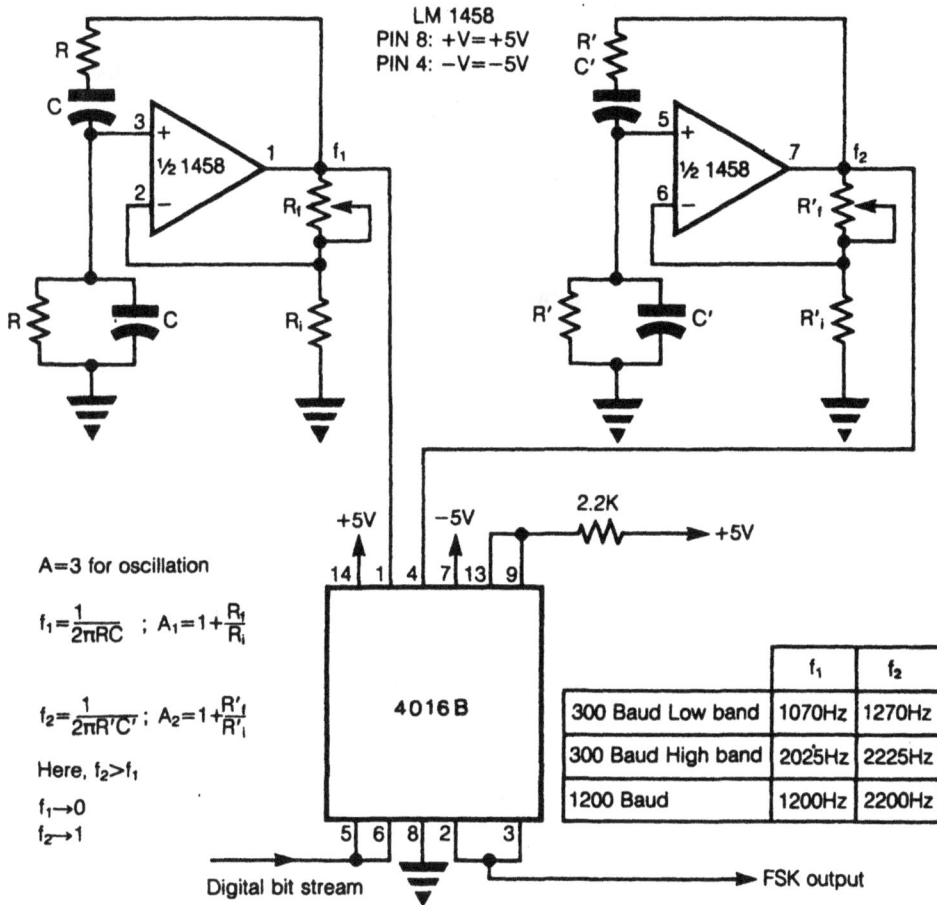

LM 1458
PIN 8: +V=+5V
PIN 4: −V=−5V

A=3 for oscillation

$$f_1 = \frac{1}{2\pi RC} \quad ; \quad A_1 = 1 + \frac{R_f}{R_i}$$

$$f_2 = \frac{1}{2\pi R'C'} \quad ; \quad A_2 = 1 + \frac{R'_f}{R'_i}$$

Here, $f_2 > f_1$

$f_1 \rightarrow 0$

$f_2 \rightarrow 1$

4016B

	f_1	f_2
300 Baud Low band	1070Hz	1270Hz
300 Baud High band	2025Hz	2225Hz
1200 Baud	1200Hz	2200Hz

Digital bit stream

FSK output

Fig. 3-9

In FSK, two discrete frequencies are used to represent the binary digits 0 and 1. The heart of the circuit consists of two Wien-bridge oscillators built using a dual op amp LM 1458, for the two frequencies. The two frequencies are enabled corresponding to digital data using two switches in SCL 4016. The control lines of these switches are logically inverted with respect to each other using one of the switches in SCL 4016 as an inverter, so as to enable only one oscillator output at a time. The digital bit stream is used to control the analog switches as shown. Since the switching frequency limit of SCL 4016 is 40 MHz, high-data rates can be easily accommodated. This method comes in handy when expensive FSK generator chips are not readily available; also, the components used in this circuit are easily available off the shelf and are quite cheap.

HARMONICS GENERATOR

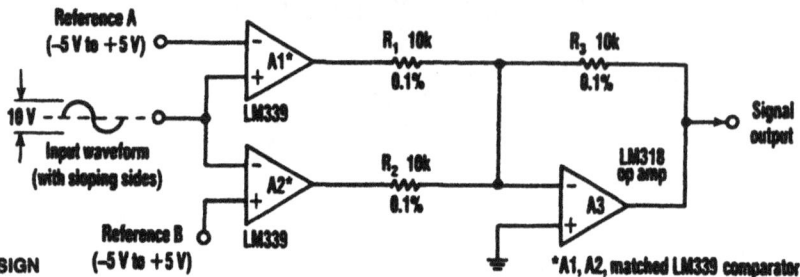

Fig. 3-10

ELECTRONIC DESIGN   *A1, A2, matched LM339 comparator

Two comparators and a summing amplifier that generate differential harmonic spectra comprise a simple frequency multiplier. The resulting circuit can extract harmonics from a sine, triangle, sawtooth, or any other sloping-sided waveform.

With a sloped-input waveform, a comparator produces an output pulse width that's proportional to the input amplitude plus a reference voltage. Changing the reference can vary the pulse width from 0 to 100%. As the pulse width changes, the harmonic spectrum changes, but two comparators combined in the adder eliminate harmonics, depending on the duty cycle. For example, a 50% pulse will lack all the even-numbered harmonics. Similarly, a 25% duty-cycle pulse will be missing multiples of the fourth harmonic and deliver the second, sixth, and tenth harmonics. Accordingly, the circuit generates multiples of the input frequency that might not have existed in the input waveform. Adjusting the references can create virtually any harmonic.

Because comparators A1 and A2 supply differential inputs to the added A3, the adder cancels out equal harmonics. Therefore, both A1 and A2 should have identical ac characteristics, and A3 should have good common-mode rejection and a high slew rate. In particular, R1, R2, and R3 should match within 0.1%. Of course, the accuracy of the circuit depends heavily on the amplitude stability of the input.

LOW-FREQUENCY FM GENERATORS

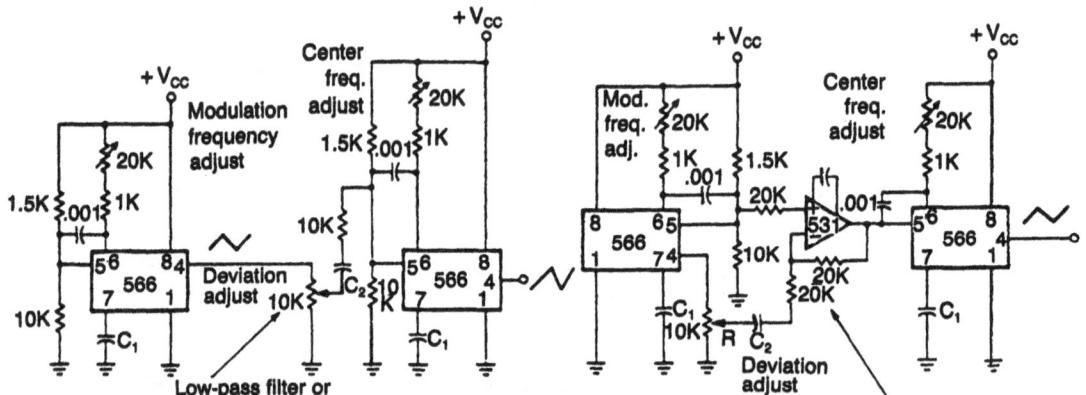

Fig. 3-11

SIGNETICS

A. Small Frequency Deviations to ± 20%

B. Large Frequency Deviations to ± 100%

Here are two FM generators for low frequency, less than 0.5-MHz center frequency, applications. Each uses a 566 function generator as a modulation generator and a second 566 as the carrier generator. Capacitor C1 selects the modulation frequency adjustment range and C1 selects the center frequency. Capacitor C2 is a coupling capacitor which only needs to be large enough to avoid distorting the modulating waveform. If a frequency sweep in only one direction is required, the 566 ramp generators given in this section can be used to drive the center generator.

POSITIVE-TRIGGERED MONOSTABLE

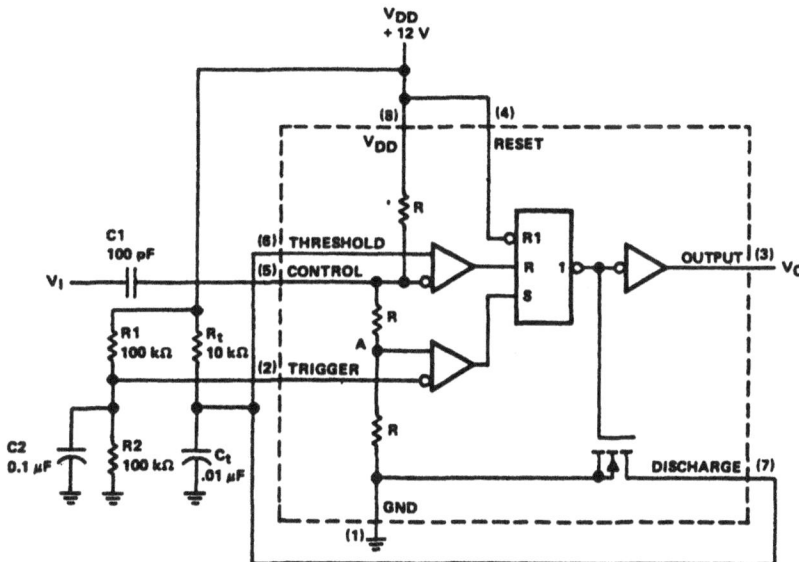

Reprinted by permission of Texas Instruments.

Fig. 3-12

A positive-going trigger pulse can be used to start the timing cycle with the circuit shown. In this design, trigger input pin 2 is biased to 6 V ($1/2$ V_{DD}) by divider R1 and R2. Control input pin 5 is biased to 8 V ($2/3$ V_{DD}) by the internal divider circuit. With no trigger voltage applied, point A is at 4 V ($1/3$ V_{DD}). To turn the timer on, the voltage at point A has to be greater than the 6 V present on pin 2. Positive 5-V trigger pulse V_I applied to the control input pin 5 is ac coupled through capacitor C1, adding the trigger voltage to the 8 V already on pin 5; this results in 13 V with respect to ground. The output pulse width is determined by the values of R_t and C_t.

When voltage at point A is increased to 6.5 V, which is greater than the 6 V on pin 2, the timer cycle is initialized. The output of timer pin 3 increases, turning off discharge transistor pin 7 and allowing C_t to charge through resistor R_t. When capacitor C_t charges to the upper threshold voltage of 8 V ($2/3$ V_{DD}), the flip-flop is reset and output pin 3 decreases. Capacitor C_t then discharges through the discharge transistor. The timer is not triggered again until another trigger pulse is applied to control input pin 5.

PRECISION AUDIO WAVEFORM GENERATOR

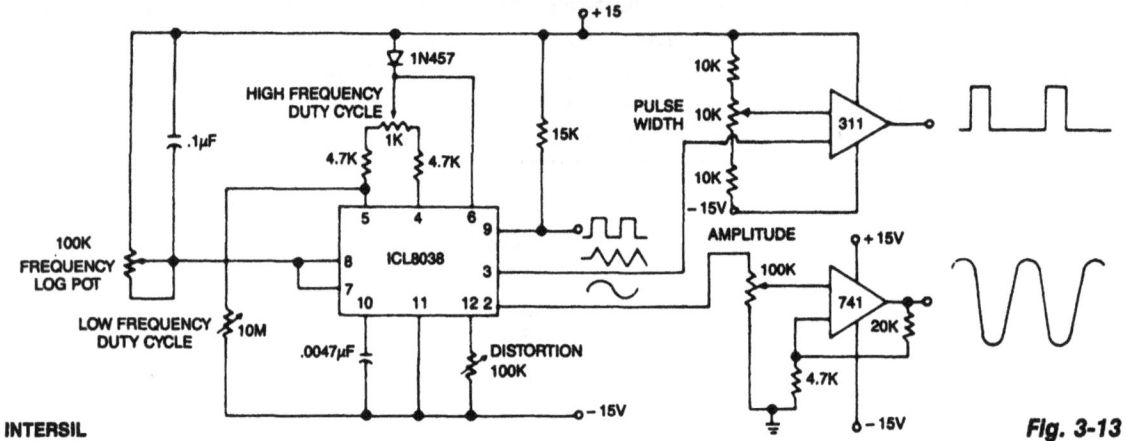

Fig. 3-13

This circuit generates sinusoidal, square, and triangle waveforms simultaneously. Set the frequency to a particular value or vary it, as shown above. An op amp can be added for extra drive capability and simplified amplitude adjustment. A simple comparator, slicing the triangle waveform, provides continuous duty cycle adjustment at a constant frequency.

MONOSTABLE MULTIVIBRATOR

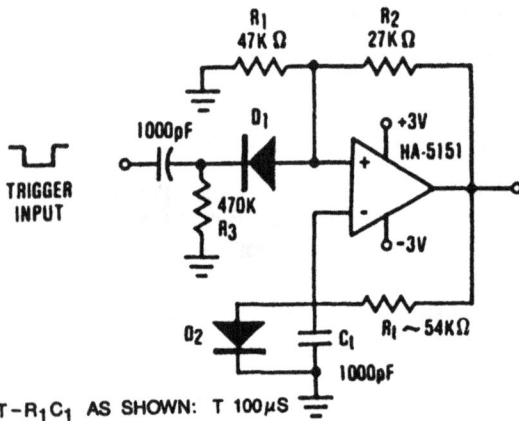

$T - R_1 C_1$ AS SHOWN: T 100μS

Fig. 3-14

The circuit illustrates the usefulness of the HA5151 as a battery-powered monostable. In this circuit, the ratio is set to 0.632, which allows the time constant equation to be reduced to:

$$T = R_t C_t$$

D2 is used to force the output to a defined state by clamping the negative input at +0.6 V. Triggering is set by C1, R3, and D2. An applied trigger pulls the positive input below the clamp voltage, +0.6 V, which causes the output to change state. This state is held because the negative input cannot follow the change because of $R_t \cdot C_t$. This particular circuit has an output pulse width set at approximately 100 μs. Use of potentiometers for R_t and variable capacitors for C_t will allow for a wide variation in T.

VERSATILE 2φ PULSE GENERATOR

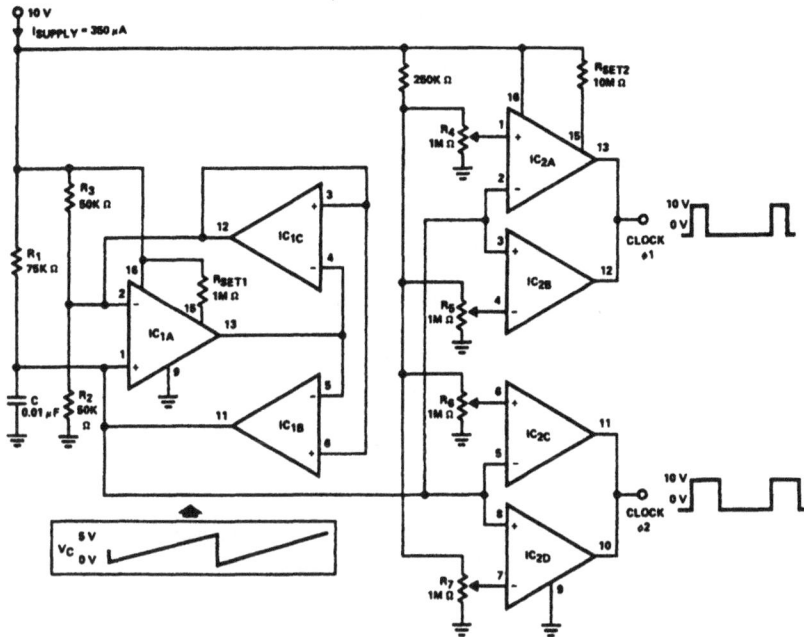

SILICONIX

Fig. 3-15

FIXED-FREQUENCY GENERATOR

A single op amp, one fourth of an LM324 quad op amp, is configured as a standard inverting amplifier. At power up, a positive voltage is applied to the noninverting input of U1, via R3, forcing its output high. That high output travels along three paths. The first path is the tone output. Along the second path, by way of R5, that high is used as the drive signal for BZ1. In the third path, the high output of U1 is fed back, via R4, to the inverting input of U1. That forces U1's output to go low. And that low, when fed back to the inverting input of U1, causes the op amp output again to a high, and the cycle repeats itself. As configured, U1 provides a voltage gain of 4.7 (gain = R4/R1).

The outer ring of the piezo element is usually connected to the circuit ground. The large inner circle serves as the driven area, and the small elongated section supplies the feedback signal. Resistor R5 sets BZ1's output-volume level. That level can be increased by decreasing R5 for example, to 470 Ω.

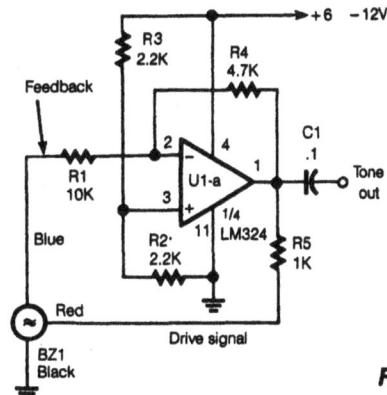

Fig. 3-16

HANDS-ON ELECTRONICS/POPULAR ELECTRONICS

Resistors R2 and R3 set the bias for op amp U1's positive input pin 3 to half of the supply-voltage level. That allows for a maximum voltage swing at U1's output. Although a quad op amp is specified, almost any similar low-cost single or dual op amp will work for U1a.

SINGLE-SUPPLY MULTIVIBRATOR

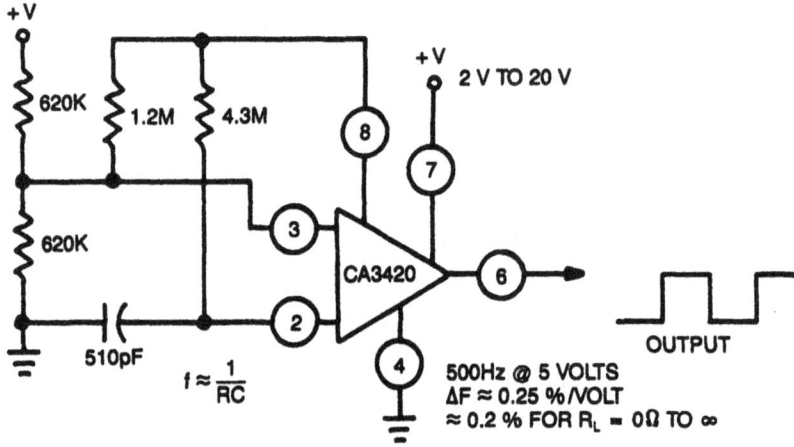

+V

620K 1.2M 4.3M

8

+V 2 V TO 20 V

7

3

CA3420

620K

6

OUTPUT

2

510pF

4

$f \approx \dfrac{1}{RC}$

500Hz @ 5 VOLTS
$\Delta F \approx 0.25\ \%/\text{VOLT}$
$\approx 0.2\ \%\ \text{FOR } R_L = 0\Omega \text{ TO } \infty$

GE/RCA

Fig. 3-17

This multivibrator uses a CA3420 BiMOS op amp to provide improved frequency stability. The output frequency remains essentially independent of supply voltage. Because of the inherent buffering action of pin 6, frequency shift is approximately 0.2% when R_L varies between zero Ω to infinity.

EASILY TUNED SINE-WAVE/SQUARE-WAVE OSCILLATOR

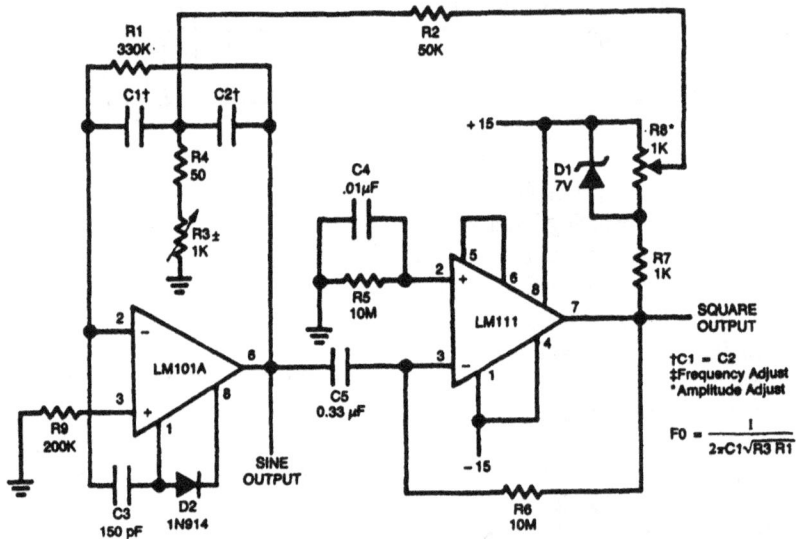

R1
330K

R2
50K

C1† C2†

+15

R8*
1K

R4
50

D1
7V

C4
.01µF

R3±
1K

R7
1K

R5
10M

5

2

LM111

6 8

7

SQUARE
OUTPUT

2

−

LM101A

6

3

4

†C1 = C2
‡Frequency Adjust
*Amplitude Adjust

3

+

8

C5
0.33 µF

1

1

$F0 = \dfrac{1}{2\pi C1\sqrt{R3\ R1}}$

R9
200K

−15

C3
150 pF

D2
1N914

SINE
OUTPUT

R6
10M

NATIONAL SEMICONDUCTOR

Fig. 3-18

84

EASILY TUNED SINE-WAVE/SQUARE WAVE OSCILLATOR (continued)

The circuit win provide both a sine- and square-wave output for frequencies from below 20 Hz to above 20 kHz. The frequency of oscillation is easily tuned by varying a single resistor. This is a considerable advantage over Wien-Bridge circuits where two elements must be tuned simultaneously to change frequency. Also, the output amplitude is relatively stable when the frequency is changed. An amp is used as a tuned circuit, driven by square wave from a voltage comparator. The frequency is controlled by R1, R2, C1, C2, and R3, with R3 used for tuning. Tuning the filter does not affect its gain or bandwidth, so the output amplitude does not change with frequency.

A comparator is fed with the sine-wave output to obtain a square wave. The square wave is then fed back to the input of the tuned circuit to cause oscillation. Zener diode, D1, stabilizes the amplitude of the square wave fed back to the filter input. Starting is ensured by R6 and C5 which provide dc negative feedback around the comparator. This keeps the comparator in the active region. Distortion ranges between 0.75% and 2% depending on the setting of R3. Although greater tuning range can be accomplished by increasing the size of R3 beyond 1 kΩ, distortion becomes excessive. Decreasing R3 lower than 50 Ω can make the filter oscillate by itself.

ASTABLE

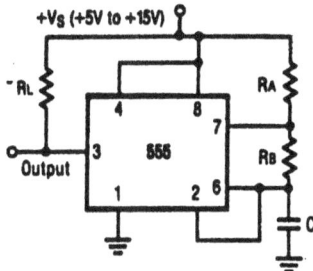

Free Running Frequency vs. R_A, R_B and C

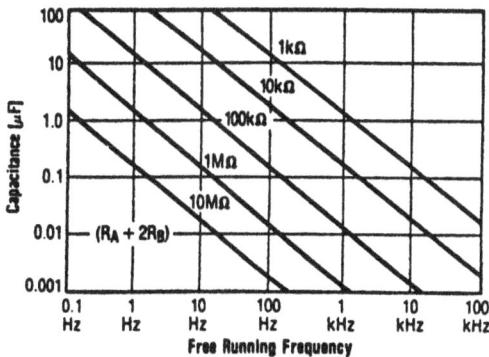

This astable will trigger itself and run free as a multivibrator. The external capacitor charges through R_A and R_B and discharges through R_B only. Thus, the duty cycle is set by the ratio of these two resistors, and the capacitor charges and discharges between $1/3$ V_S and $2/3$ V_S. The charge and discharge times, and therefore frequency, are independent of supply voltage. The free-running frequency versus R_A, R_B and C is shown in the graph.

Reprinted with permission from Raytheon Co., Semiconductor Division. *Fig. 3-19*

TWO-FUNCTION SIGNAL GENERATOR

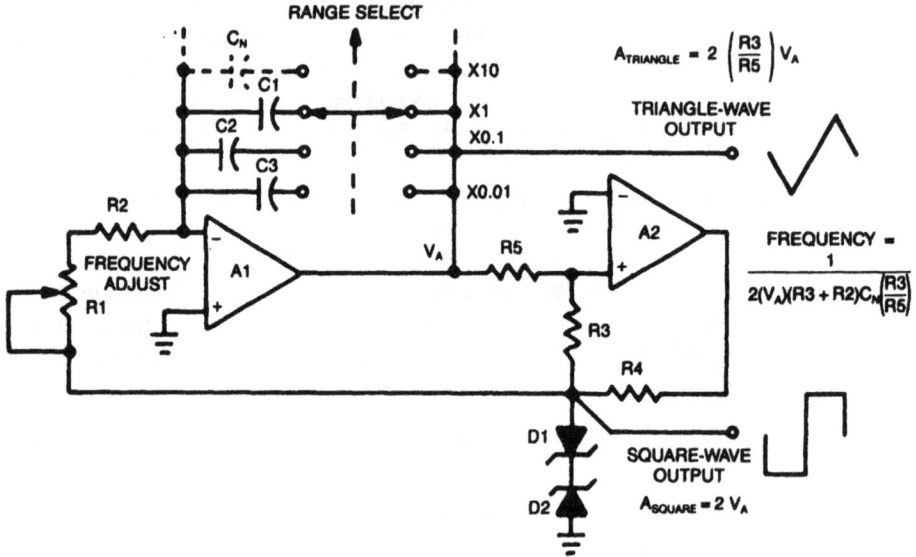

$$A_{TRIANGLE} = 2 \left(\frac{R3}{R5} \right) V_A$$

TRIANGLE-WAVE OUTPUT

$$FREQUENCY = \frac{1}{2(V_A)(R3 + R2)C_N \left(\frac{R3}{R5} \right)}$$

SQUARE-WAVE OUTPUT

$$A_{SQUARE} = 2 V_A$$

EDN

Fig. 3-20

You can continuously vary the frequencies of the triangle and square waves produced by this circuit over a full decade. If R5 = R3, the amplitude of the two waveforms will be equal ($A_{SQUARE} = A_{TRIANGLE}$).

TRIANGLE GENERATOR

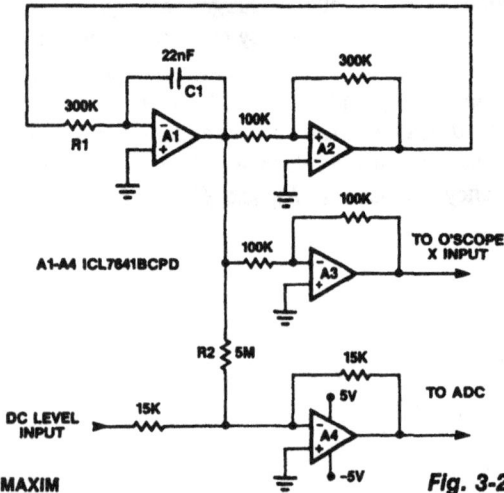

A1-A4 ICL7641BCPD

MAXIM

Fig. 3-21

This circuit generates a symmetrical, 10-mV pk-pk triangle waveform which is summed with a dc level and connected to the a/d analog input for noise/DNL testing. The dc level input offsets the triangle waveform over the input range of the ADC. The 10-mV amplitude amounts to an 8 LSB span for a 12-bit, 5-V, full-scale ADC.

MONOSTABLE

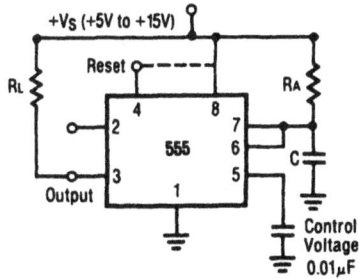

Time Delay vs. R_A, R_B and C

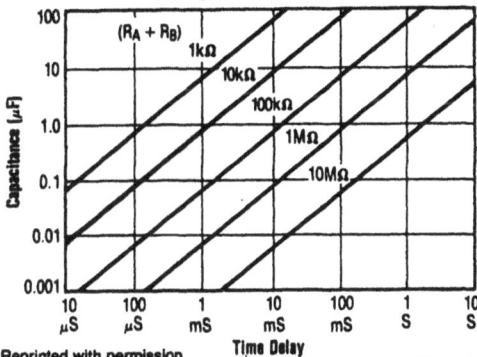

Reprinted with permission
from Raytheon Co., Semiconductor Division.

Fig. 3-22

In this mode, the timer functions as a one shot. The external capacitor is initially held discharged by a transistor internal to the timer. Applying a negative trigger pulse to pin 2 sets the flip-flop, driving the output high, and releasing the short circuit across the external capacitor. The voltage across the capacitor increases with the time constant $r = R_A C$ to $2/3$ V_S, where the comparator resets the flip-flop and discharges the external capacitor. The output is now in the low state.

Circuit triggering takes place when the negative-going trigger pulse reaches $1/3$ V_S; the circuit stays in the output high state until the set time elapses. The time the output remains in the high state is 1.1 $R_A C$ and can be determined by the graph. A negative pulse applied to pin 4 (reset) during the timing cycle will discharge the external capacitor and start the cycle over again beginning on the positive-going edge of the reset pulse. If reset function is not used, pin 4 should be connected to V_S to avoid false resetting.

PROGRAMMABLE-FREQUENCY, FREE-RUNNING MULTIVIBRATOR

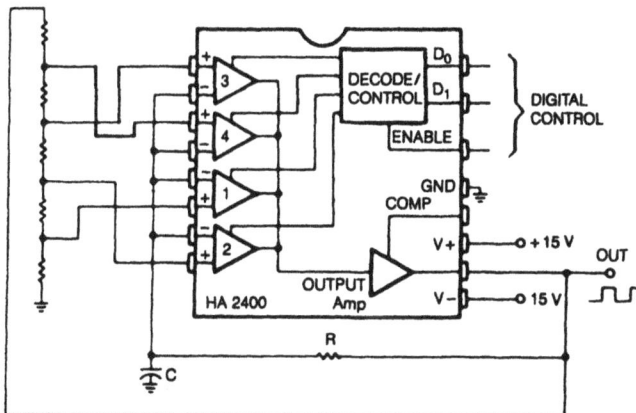

HARRIS

This is the simplest of any programmable oscillator circuit, since only one stable timing capacitor is required. The output square wave is about 25 V pk-pk, and has rise and fall times of about 0.5 μs. If a programmable attenuator circuit is placed between the output and the divider network, 16 frequencies can be produced with two HA-200's and still only one timing capacitor.

87

FUNCTION GENERATOR

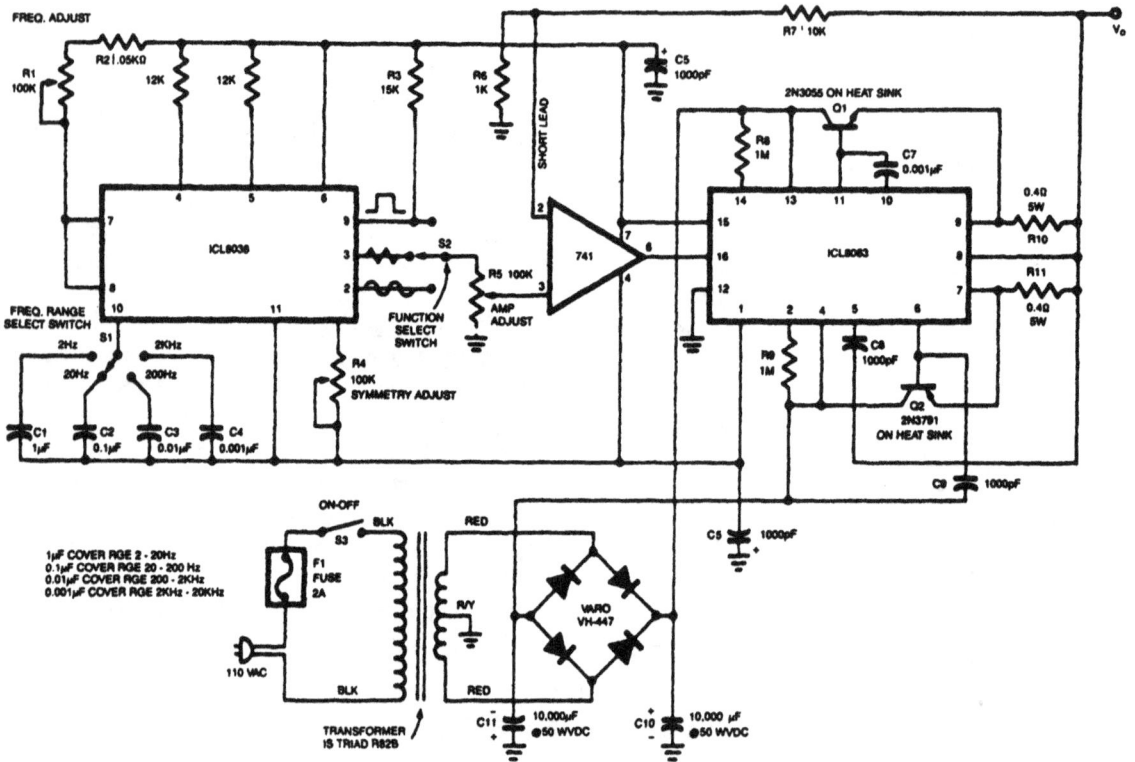

INTERSIL

Fig. 3-24

This generator will supply sine, triangular, and square waves from 2 Hz to 20 kHz. This complete test instrument can be plugged into a standard 110 Vac line for power. V_{OUT} will be up to ± 25 V (50 V pk-pk across loads as small as 10 Ω (about 2.5 A maximum output current).

Capacitor working voltages should be greater than 50 Vdc and all resistors should be 1/2 W, unless otherwise indicated. The interconnecting leads from the 741 pins 2 and 3 to their respective resistors should be kept short, less than 2 inches if possible; longer leads might result in oscillation. Full output swing is possible to about 5 kHz; after that the output begins to taper off because of the slew rate of the 741, until at 20 kHz the output swing will be about 20 V_{pp} ± 10 V. This problem can be remedied by simply using an op amp with a higher slew rate, such as the LF356.

PROGRAMMABLE-FREQUENCY ASTABLE

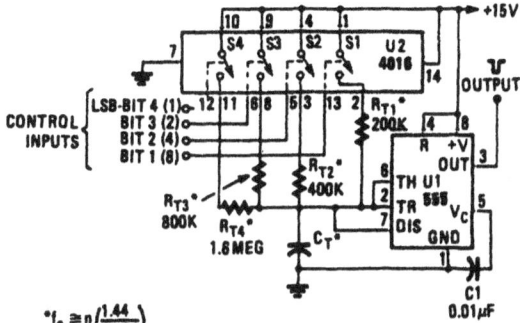

$$^*f_o \cong n\left(\frac{1.44}{R_T C_T}\right)$$

WHERE,

n is DIGITAL INPUT WORD: $1 < n < 15$
(AS SHOWN, WITH BASE R_T OF 1.6 MEG, 100 Hz $< f_o$ 1500 Hz).

POPULAR ELECTRONICS *Fig. 3-25*

LINEAR-RAMP MONOSTABLE

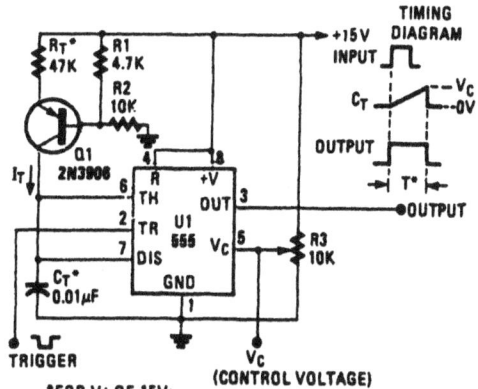

*FOR V+ OF 15V:

$$T = \frac{V_C C_T}{I_T}, \quad I_T \cong \frac{4.2}{R_T} \qquad T \cong 0.24 V_C R_T C_T$$

(AS SHOWN, $T_{MAX} \cong 1$ MS WITH $V_C = 10$V.)

HANDS-ON ELECTRONICS *Fig. 3-26*

LOW-FREQUENCY MULTIVIBRATOR

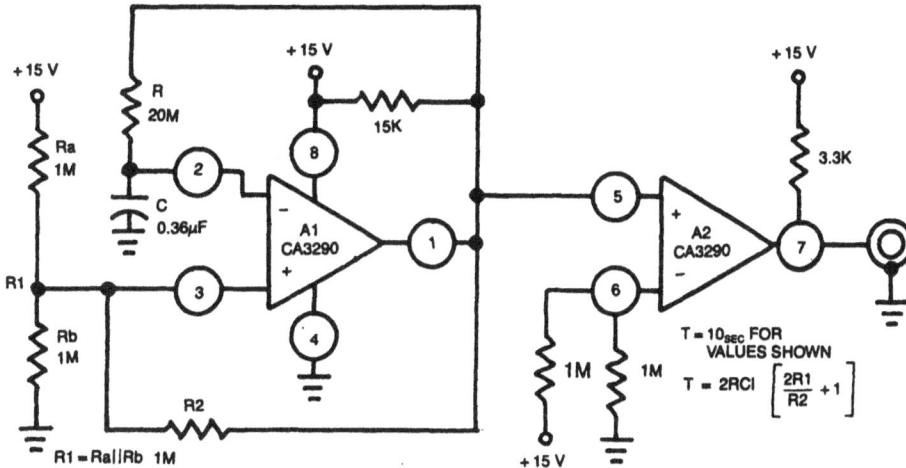

$$T = 2RCl\left[\frac{2R1}{R2} + 1\right]$$

T = 10 SEC FOR VALUES SHOWN

ALL RESISTANCE VALUES ARE IN OHMS

GE/RCA *Fig. 3-27*

This circuit uses half of the CA3290 BiMOS dual voltage comparator as a conventional multivibrator. The second half maintains frequency against effects of output loading. Large values of timing resistor, R1, ensure long time delays with low-leakage capacitors.

RETRIGGERABLE ONE SHOT

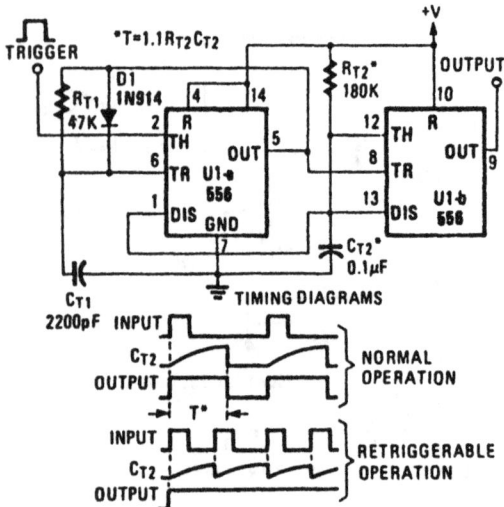

HANDS-ON ELECTRONICS

Fig. 3-28

ASTABLE MULTIVIBRATOR

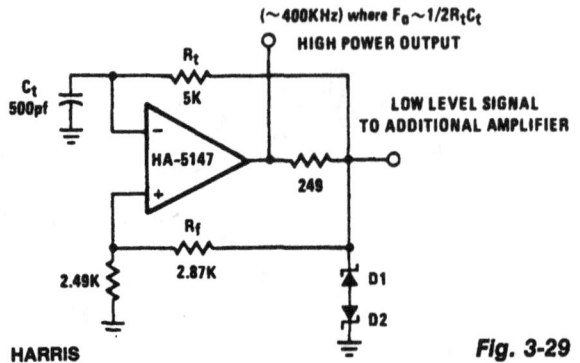

$(\sim 400KHz)$ where $F_0 \sim 1/2R_tC_t$

HIGH POWER OUTPUT

LOW LEVEL SIGNAL
TO ADDITIONAL AMPLIFIER

HARRIS

Fig. 3-29

The power bandwidth of the HA-5147 extends the circuit's frequency range to approximately 500 kHz. R_t can be made adjustable to vary the frequency if desired. Any timing errors because of V_{OS} or I_{bias} have been minimized by the precision characteristics of the HA-5147. D1 and D2, if used, should be matched to prevent additional timing errors. These clamping diodes can be omitted by tying R_t, and positive feedback resistor R_f directly to the output.

SINGLE-CONTROL FUNCTION GENERATOR

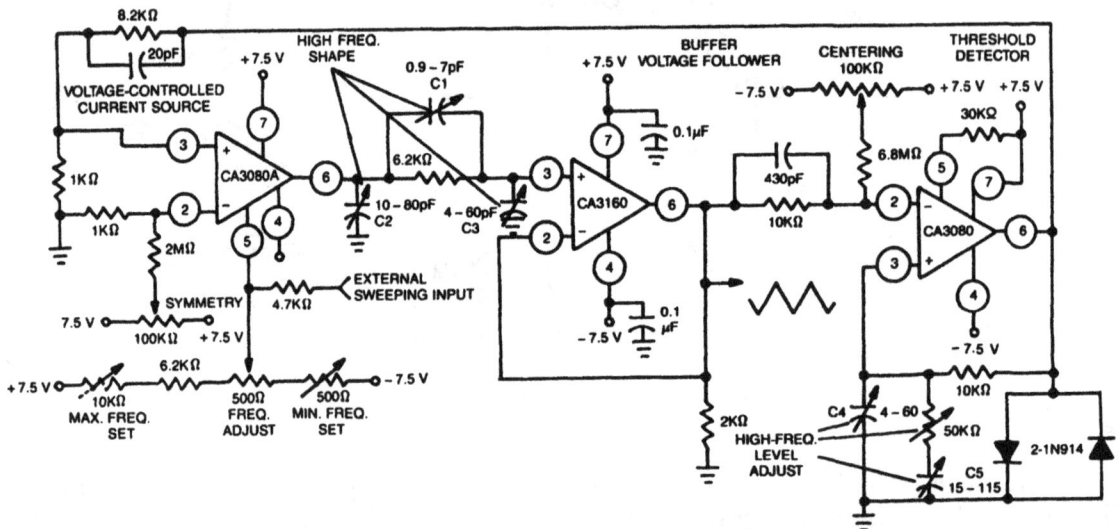

GE/RCA

Fig. 3-30

SINGLE-CONTROL FUNCTION GENERATOR (continued)

This function generator, with an adjustment range in excess of 1,000,000 to 1, uses a CA3160 BiMOS op amp as a voltage follower, a CA3080 OTA as a high-speed comparator, and a CA3080 as a programmable-current source. Three variable capacitors, C1, C2, and C3 shape the triangular signal between 500 kH and 1 MHz. Capacitors C4 and C5, and the trimmer potentiometer in series with C5, maintain essentially constant ($\pm 10\%$) amplitude to 1 MHz.

TRIANGLE-/SQUARE-WAVE GENERATOR

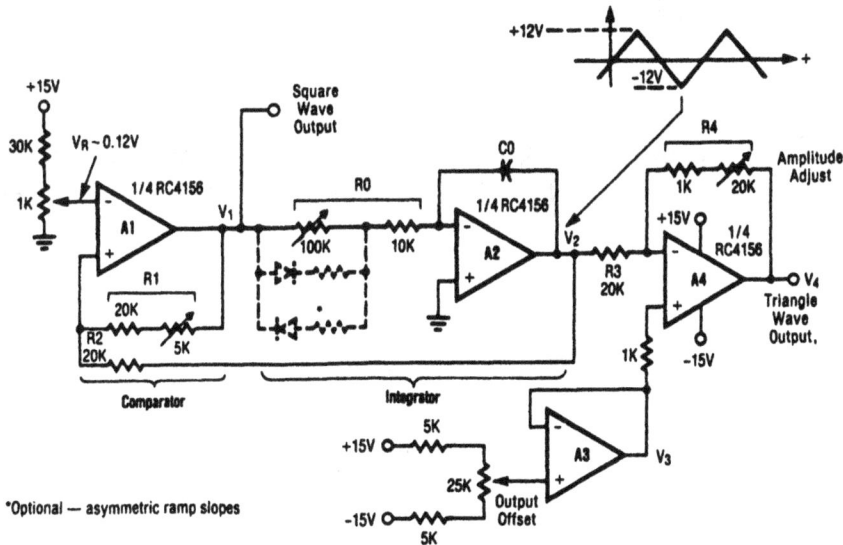

Reprinted with permission from Raytheon Co., Semiconductor Division.

Fig. 3-31

This circuit uses a positive-feedback loop closed around a combined comparator and integrator. When power is applied, the output of the comparator will switch to one of two states, to the maximum positive or maximum negative voltage. This applies a peak input signal to the integrator, and the integrator output will ramp either down or up, opposite of the input signal. When the integrator output, which is connected to the comparator input, reaches a threshold set by R1 and R2, the comparator will switch to the opposite polarity. This cycle will repeat endlessly, the integrator charging positive then negative, and the comparator switching in a square-wave fashion.

VARIABLE DUTY CYCLE TIMER

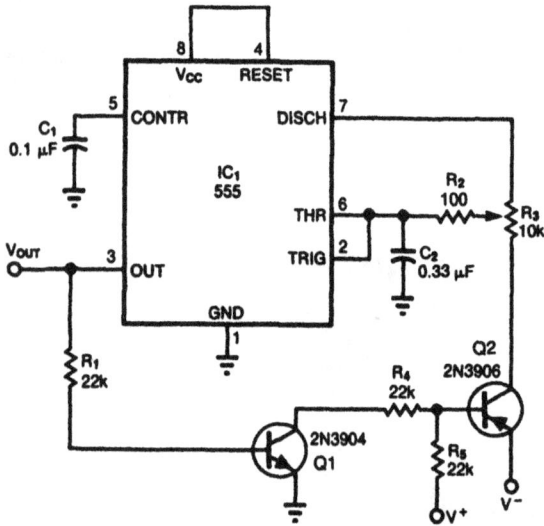

When configured as a free-running multivibrator, a 555 timer provides no more than a 50% duty cycle. By adding two transistors, however, you can obtain a variable 5 to 95% duty cycle without changing the sum of the on and off times. When V_{OUT} decreases, Q1 is on and Q2 is off, disconnecting V+ while timing capacitor C2 discharges into pin 7 of the timer. When V_{OUT} increases, Q2 reconnects V+ for recharging C2.

Adjusting linear trimming potentiometer R3 to increase the charging resistance increases the on time, but decreases the off time by the same amount by lowering the discharge resistance (the converse is also true). As a result, the sum of the on and off times remains constant. R2 protects Q2 and the timer against high-charge/discharge currents.

EDN   Fig. 3-32

BASIC FUNCTION GENERATOR

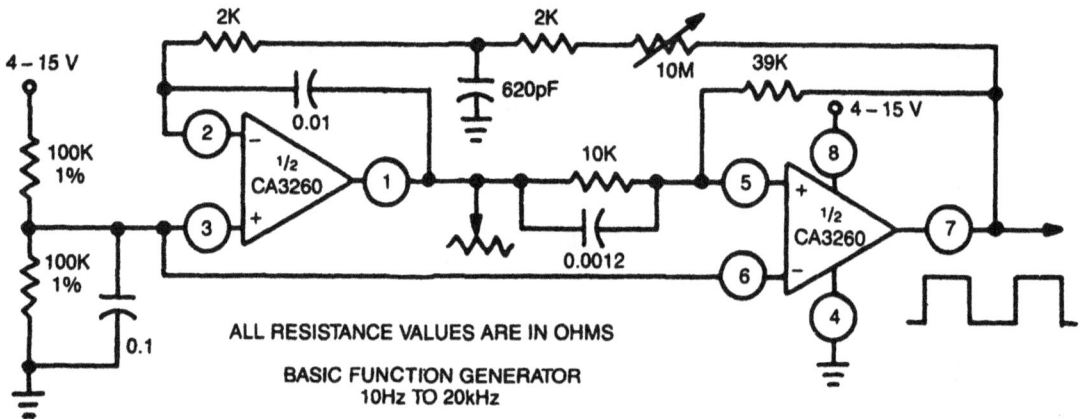

ALL RESISTANCE VALUES ARE IN OHMS

BASIC FUNCTION GENERATOR
10Hz TO 20kHz

GE/RCA   Fig. 3-33

This function generator uses a CA3260 BiMOS op amp to perform both the integrator and switching functions. A 620-pF capacitor and 2-kΩ resistor shape feedback square wave to reduce spikes. Full audio spectrum, 10 Hz to 20 kHz, is covered with a single 10-Ω potentiometer. This circuit requires 9-V battery.

WIDE-RANGE TUNABLE FUNCTION GENERATOR

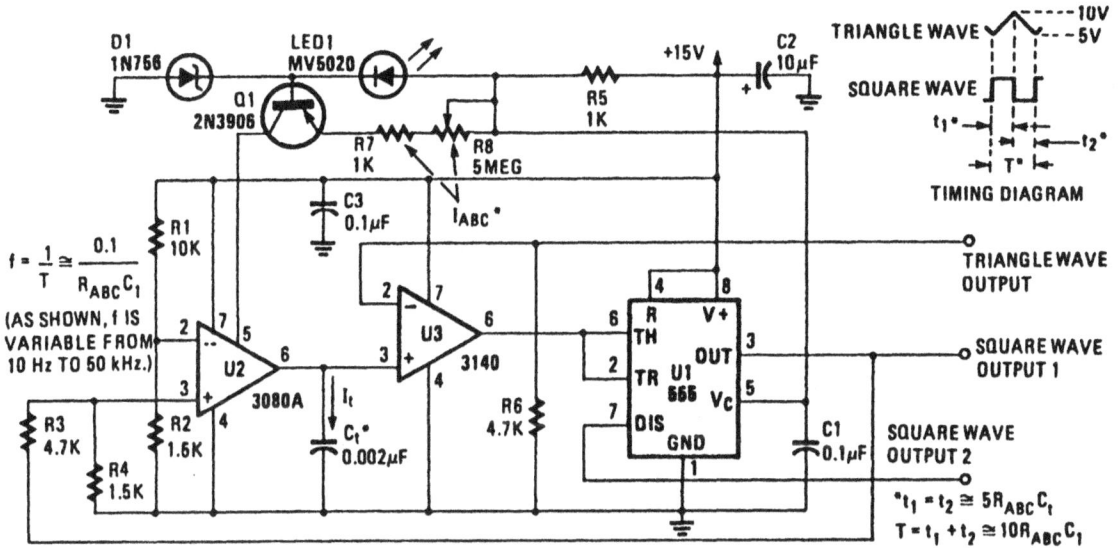

Fig. 3-34

SAWTOOTH AND PULSE GENERATOR

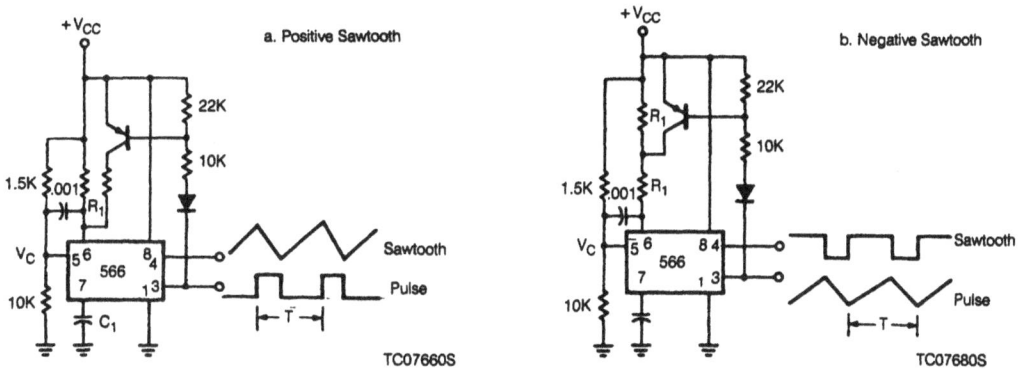

a. Positive Sawtooth

b. Negative Sawtooth

Fig. 3-35

The pin 3 output of the 566 can be used to provide different charge and discharge currents for C1 so that a sawtooth output is available at pin 4 and a pulse at pin 3. The pnp transistor should be well saturated to preserve good temperature stability. The charge and discharge times can be estimated by using the formula:

$$T = \frac{R_T C_I V_{CC}}{5(V_{CC} - V_C)}$$

where R_T is the combined resistance between pin 6 and V_{CC} for the interval considered.

SINE/COSINE GENERATOR (0.1 to 10 kHz)

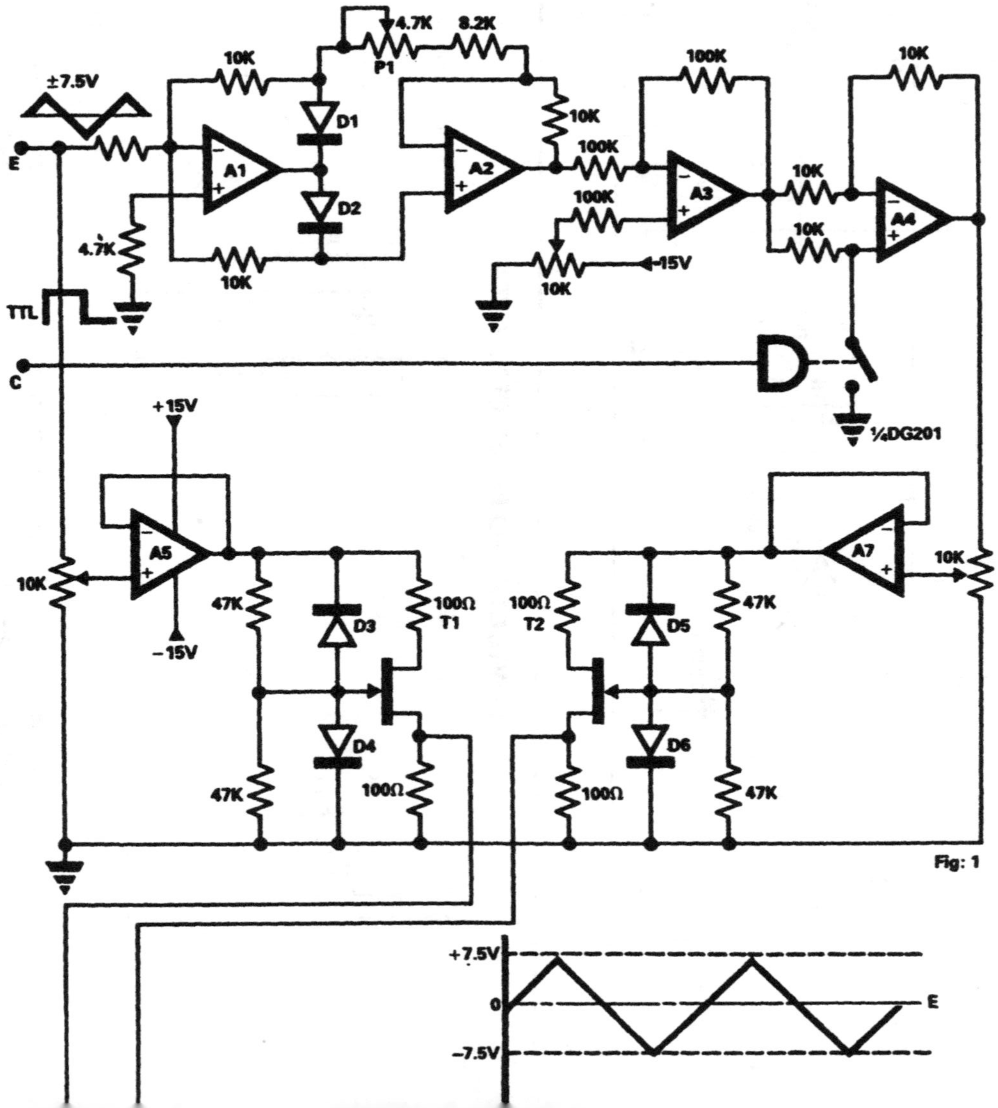

Fig: 1

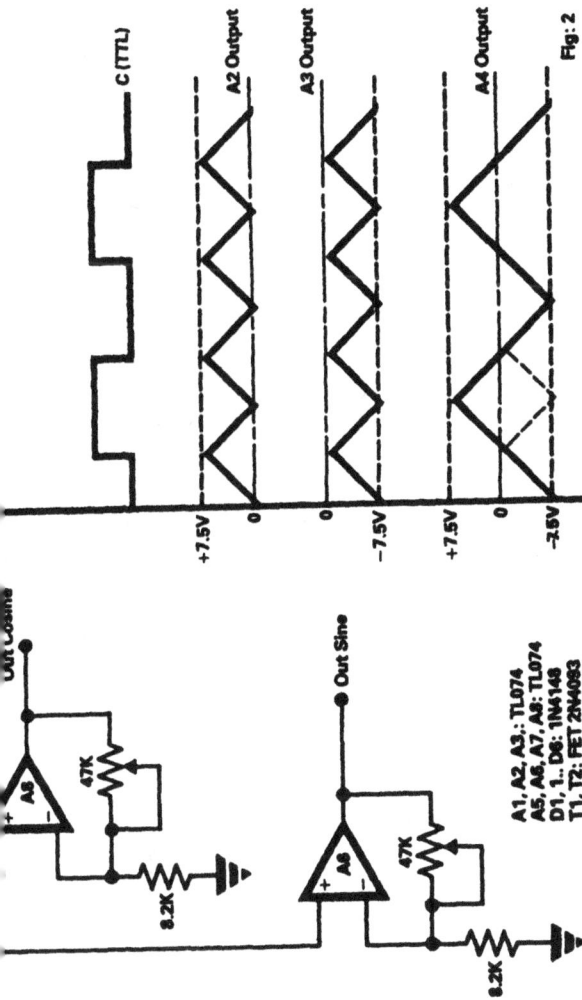

Fig. 3-36

The scheme presented delivers waveforms from any function generator producing a triangular output and a synchronized TTL square wave. A1 and A2 act as a two-phase current rectifier by inverting the negative voltage appearing at the input of A1.

Positive input: Both A1 and A2 work as unity-gain followers, D1 and D2 being in the off-state.

Negative input: A1 has a $-2/3$ gain (D1 off and D2 on), A2 has a $+3/2$ gain and the total voltage transfer is -1 between output and input. P1 allows a fine trimming of the -1 gain for the negative input signals. A3 adds a continuous voltage to the rectified positive signal in order to attack A4 which acts as a $\pm$ multiplier commanded by the TTL input through the analog switch. The signal polarity is reconstructed and the output of A4 delivers a triangular waveform, shifted by 90°, with respect to the input signal (Fig. 2). The original and the shifted voltages are fed into the triangle-to-sine converters through A5 and A7, working as impedance converters. Over the frequency dynamic ranges, from 0.1 Hz to 10 kHz, the phase shift is constant and the distortion on the sine voltage is less than 1%.

QUAD OP AMP SYNCHRONIZED FUNCTION GENERATOR

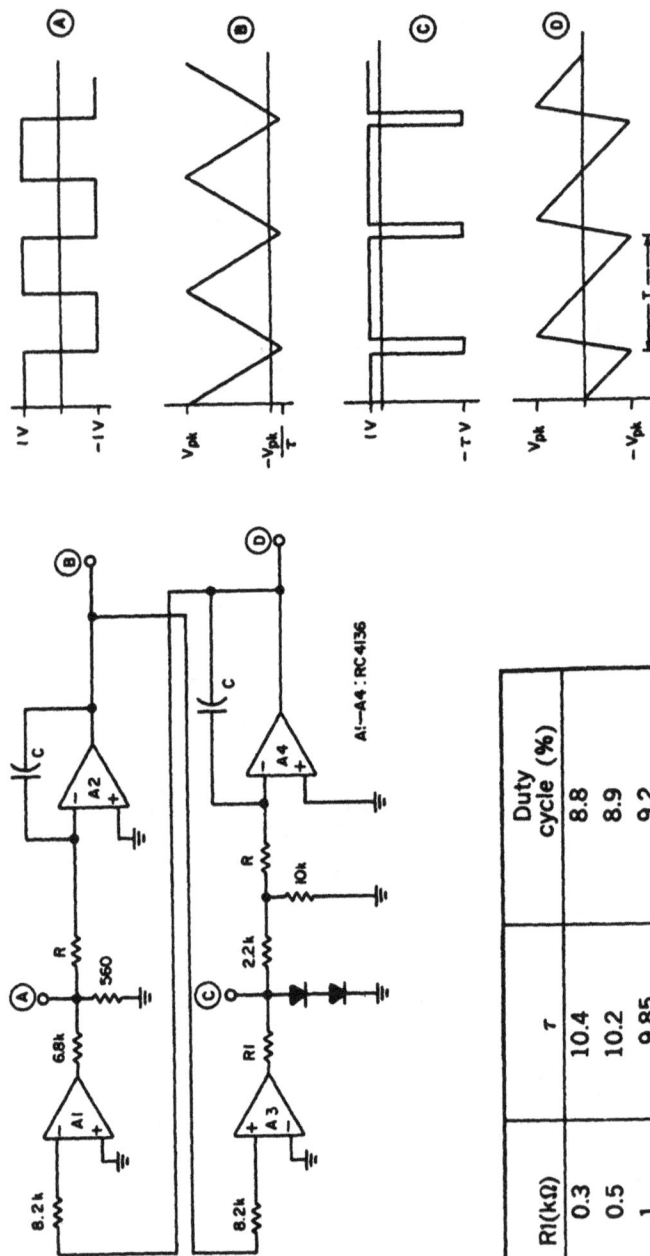

A1–A4 : RC 4136

R1(kΩ)	τ	Duty cycle (%)
0.3	10.4	8.8
0.5	10.2	8.9
1	9.85	9.2
4	8.0	11.1
10	5.86	14.6
20	4.04	19.84
50	2.09	32.4
100	1.16	46.30
117.8	1	50

ELECTRONIC DESIGN

Fig. 3-37

A quad op amp can simultaneously generate four synchronized waveforms. The two comparators (A1 and A3) produce square and pulse waves, while the two integrators (A2 and A4) give triangular and sawtooth waves. Resistor R1 sets the duty cycle and the frequency, along with resistors R and capacitors C.

OSCILLATOR OR AMPLIFIER WITH WIDE FREQUENCY RANGE

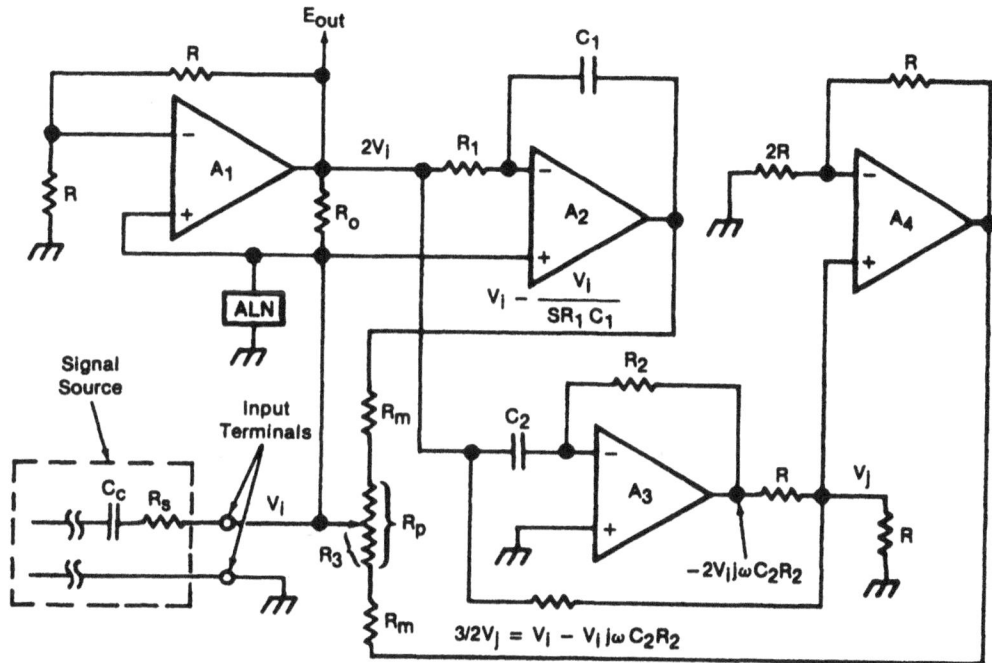

NOTES: 1. A_1, A_2, A_3, and A_4 are operational amplifiers
2. ALN = Amplitude-Limiting Network

NASA

Fig. 3-38

An oscillator/amplifier is resistively tunable over a wide frequency range. Feedback circuits containing operational amplifiers, resistors, and capacitors synthesize the electrical effects of an inductance and capacitance in parallel between the input terminals. The synthetic inductance and capacitance, and, therefore, the resonant frequency of the input admittance, are adjusted by changing a potentiometer setting. The input signal is introduced in parallel to the noninverting input terminals of operational amplifiers A1 and A2 and to the potentiometer cursor. The voltages produced by the feedback circuits, in response to input voltage, V_1 are indicated at the various circuit nodes.

LINEAR TRIANGLE-/SQUARE-WAVE VCO

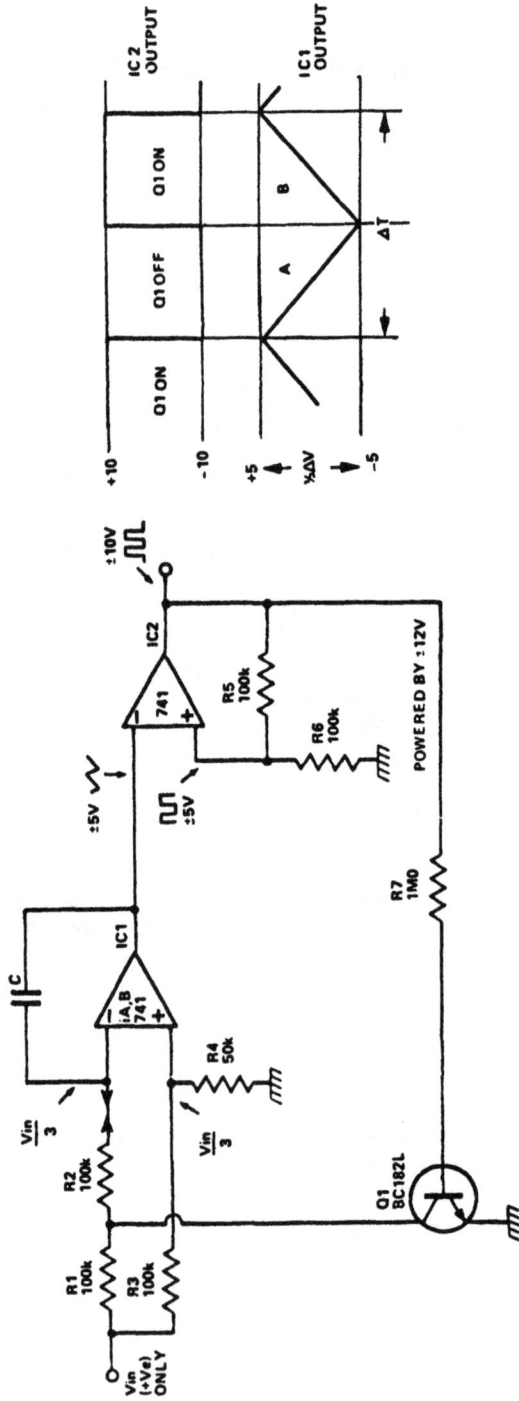

ELECTRONICS TODAY INTERNATIONAL

Fig. 3-39

The VCO has two buffered outputs; a triangle wave and a square wave. The frequency is dependent on the output voltage swing of the Schmitt trigger, IC2. Superior performance can be obtained by replacing Q1 with a switching FET. Fast FET op amps will improve high-frequency performance.

CIRCUIT FOR MULTIPLYING PULSE WIDTHS

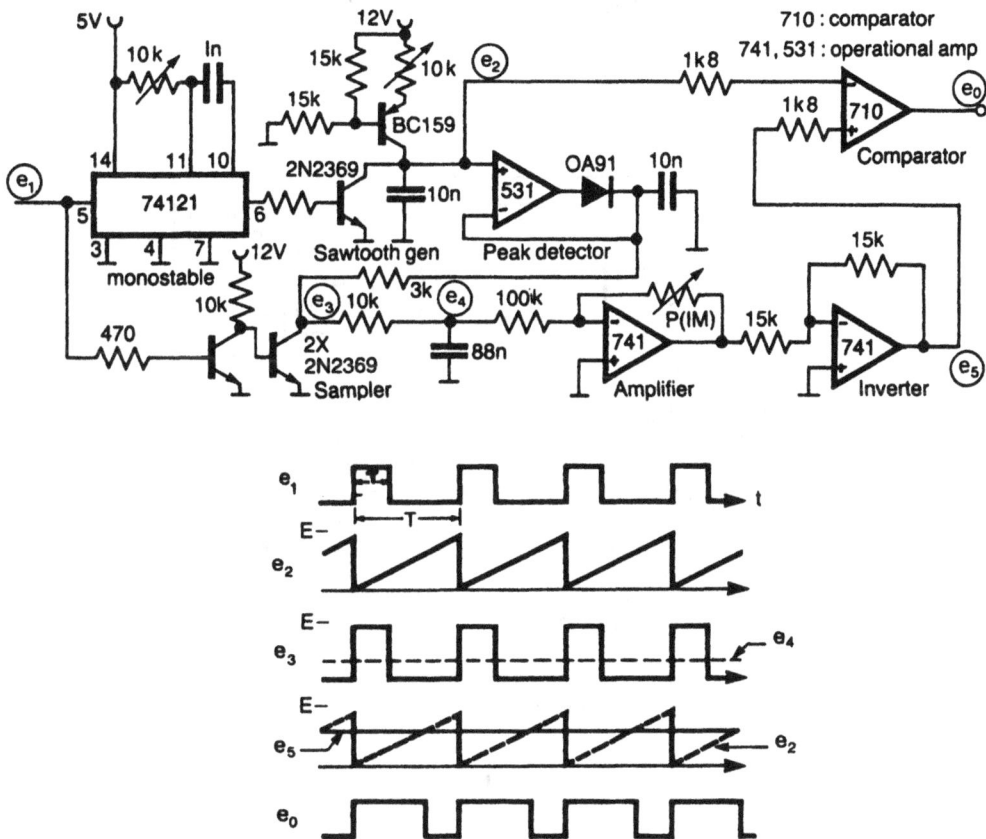

Fig. 3-40

A circuit for multiplying the width of incoming pulses by a factor of greater or less than unity is simple to build. Also, the multiplying factor can be selected by adjusting one potentiometer only. The multiplying factor is determined by setting the potentiometer P in the feedback of a 741 amplifier. The input pulses e_1 of width τ and repetition period T is used to trigger a sawtooth generator at its rising edges to produce the waveform e_2 having a peak value of (E) volt. This peak value is then sampled by the input pulses to generate the pulse train e_3 having an average value of $e_4 (= E\ E\tau/T)$, which is proportional to τ and independent on T. The dc voltage e_4 is amplified by a factor k and compared with sawtooth waveform e_2 giving output pulses of duration $k\ \tau$. The circuit is capable of operating over the frequency range 10 to 100 kHz.

PROGRAMMABLE VOLTAGE-CONTROLLED FREQUENCY SYNTHESIZER

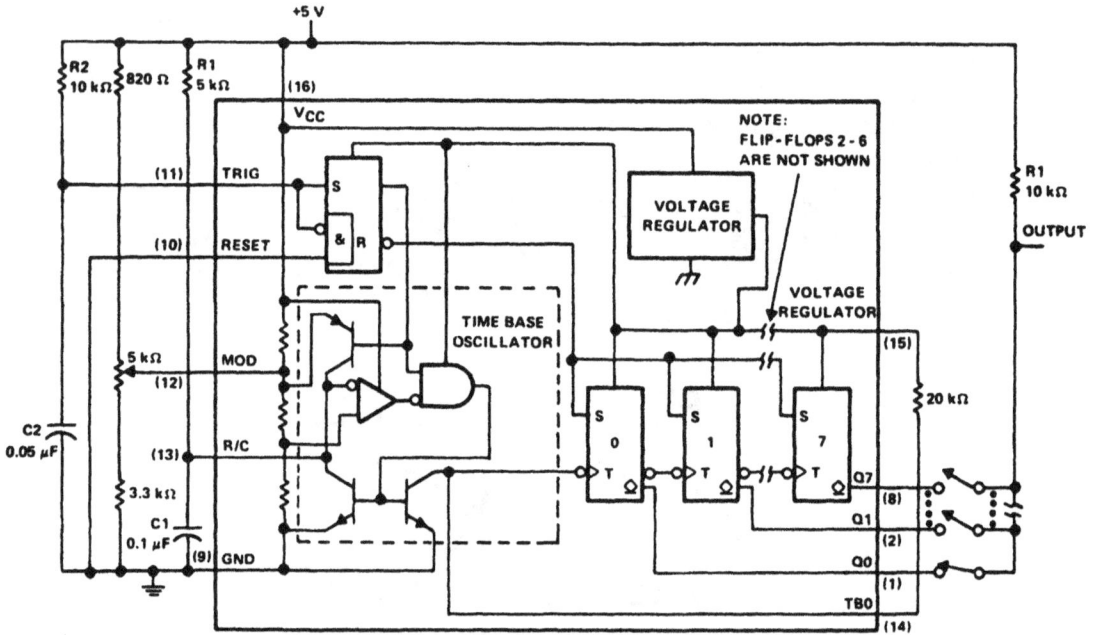

TEXAS INSTRUMENTS

Fig. 3-41

The μA2240 consists of four basic circuit elements: (1) a time-base oscillator, (2) an eight-bit counter, (3) a control flip-flop, and (4) a voltage regulator. The basic frequency of the time-base oscillator (TBO) is set by the external time constant, which is determined by the values of R1 and C1 (1?R1C1 = 2 kHz). The open-collector output of the TBO is connected to the regulator output via a 20-kΩ pull-up resistor, and drives the input to the eight-bit counter. At power-up, a positive trigger pulse is detected across C2 which starts the TBO and sets all counter outputs to a low state. Once the μA2240 is initially triggered, any further trigger inputs are ignored until it is reset. In this astable operation, the μA2240 will free-run from the time it is triggered until it receives an external reset signal. Up to 255 discrete frequencies can be synthesized by connecting different counter outputs.

EMITTER-COUPLED RC OSCILLATOR

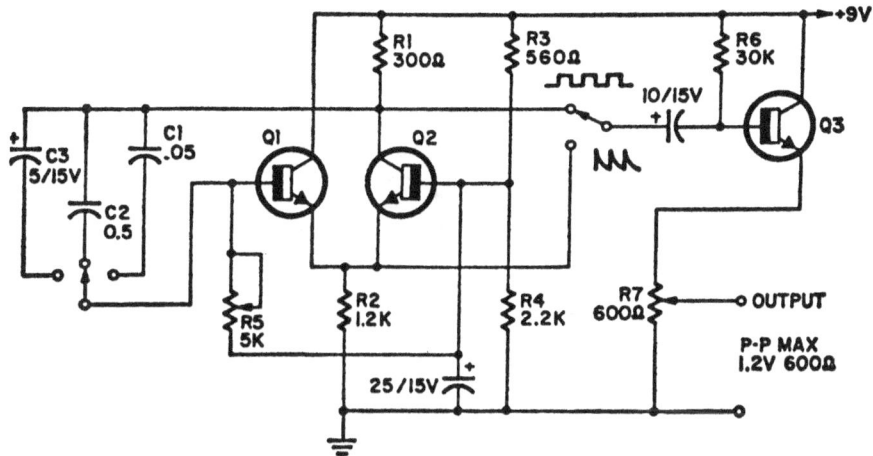

RADIO-ELECTRONICS

Fig. 3-42

The circuit covers 15 Hz to 30 kHz and is useful as a function generator. The 2N2926 or equivalent transistors can be used.

VOLTAGE-CONTROLLED HIGH-SPEED ONE SHOT

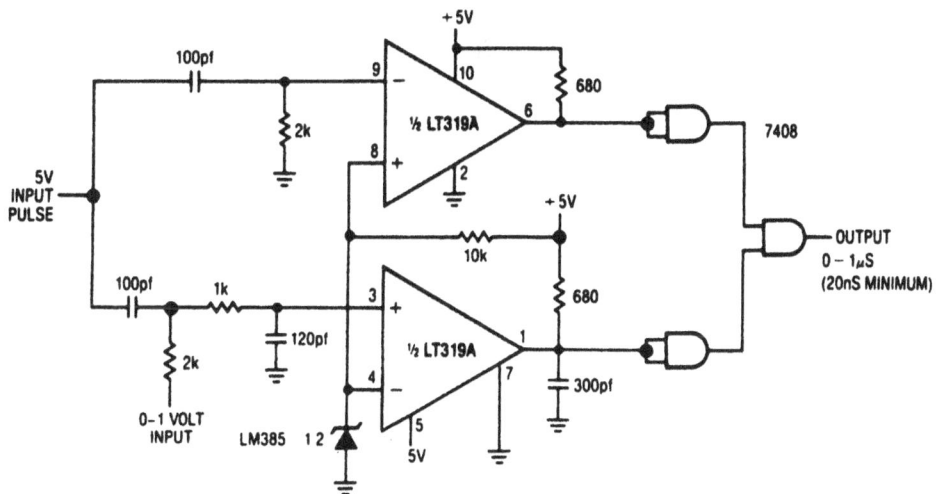

LINEAR TECHNOLOGY CORP.

Fig. 3-43

RAMP GENERATOR WITH VARIABLE RESET LEVEL

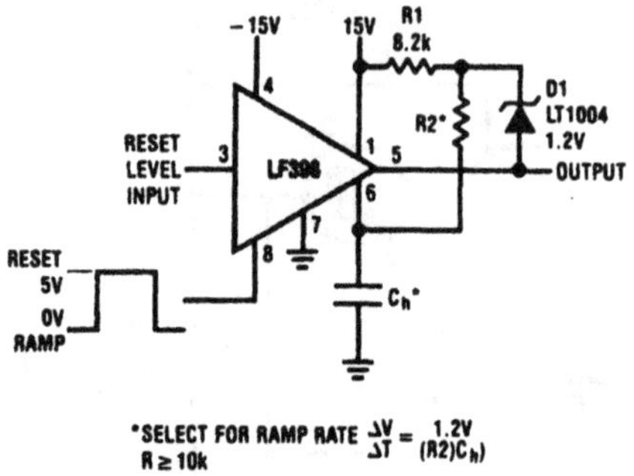

Fig. 3-44

*SELECT FOR RAMP RATE $\frac{\Delta V}{\Delta T} = \frac{1.2V}{(R2)C_h)}$
$R \geq 10k$

LINEAR TECHNOLOGY CORP.

555 ASTABLE WITH LOW DUTY CYCLE

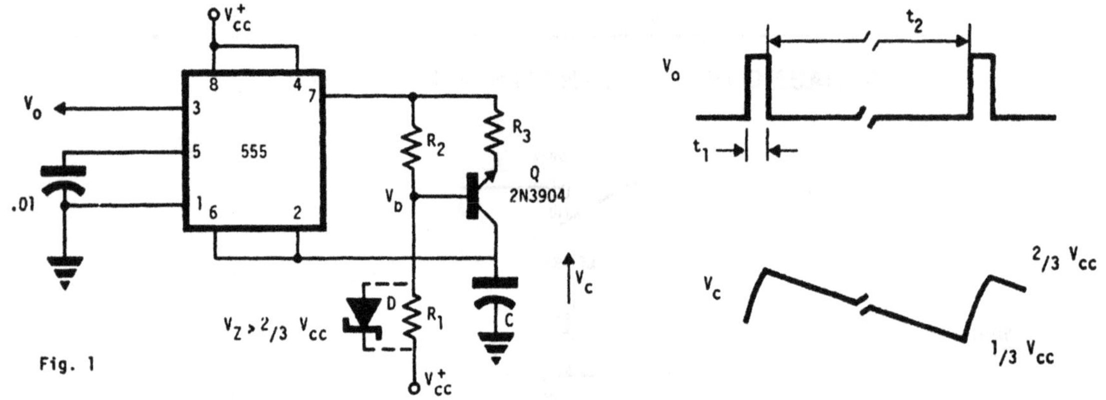

Fig. 1

$V_Z > \frac{2}{3} V_{CC}$

ELECTRONIC ENGINEERING

Fig. 3-45

This free-running multivibrator uses an external current sink to discharge the timing capacitor, C. Therefore, interval t_2 can easily be 1000 × the pulse duration, t_1, which defines a positive output. Capacitor voltage, V_C, a negative-going ramp with exponential rise during the pulse output periods.

MONOSTABLE USING VIDEO AMPLIFIER AND COMPARATOR

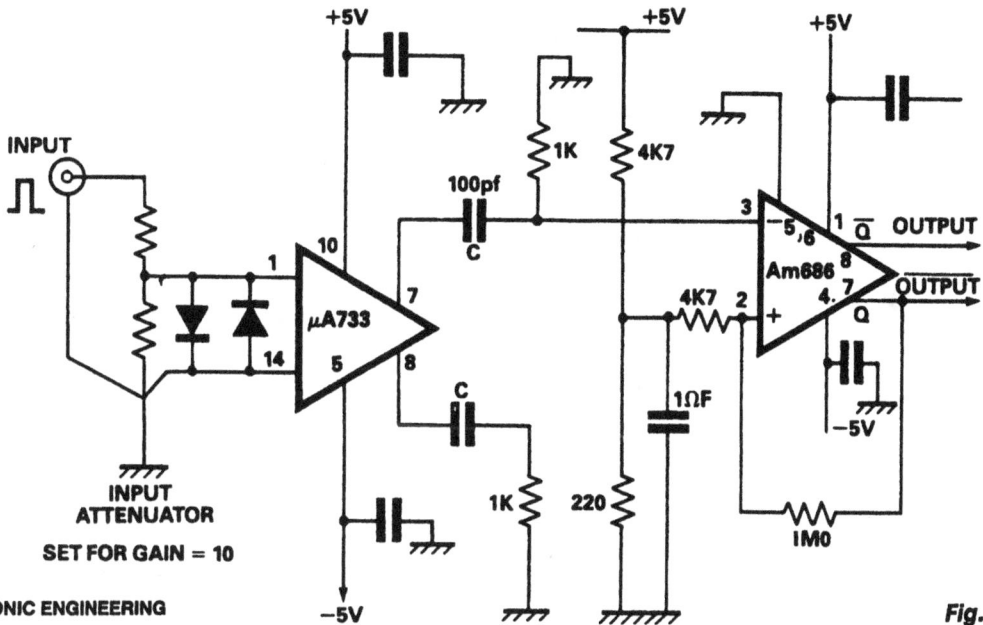

ELECTRONIC ENGINEERING

Fig. 3-46

The output of a video amplifier is differentiated before being fed to a Schottky comparator. The propagation delay is reduced to typically 10 ns. The output pulse width is set by the value of C; 100 pf gives a pulse of about 90 ns duration.

UJT MONOSTABLE CIRCUIT INSENSITIVE TO CHANGE IN BIAS VOLTAGE

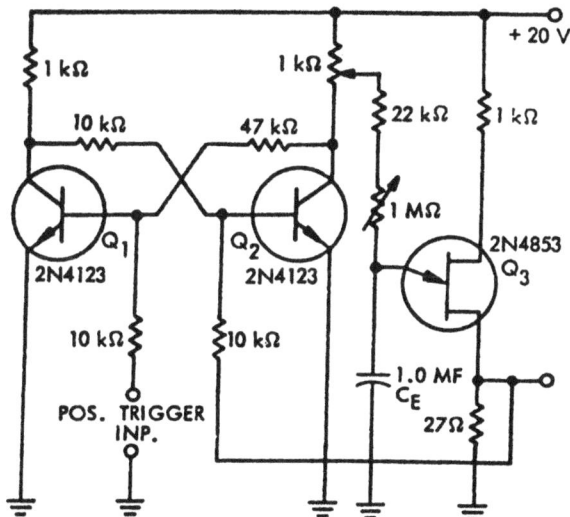

MOTOROLA   Fig. 3-47

LINEAR RAMP GENERATOR

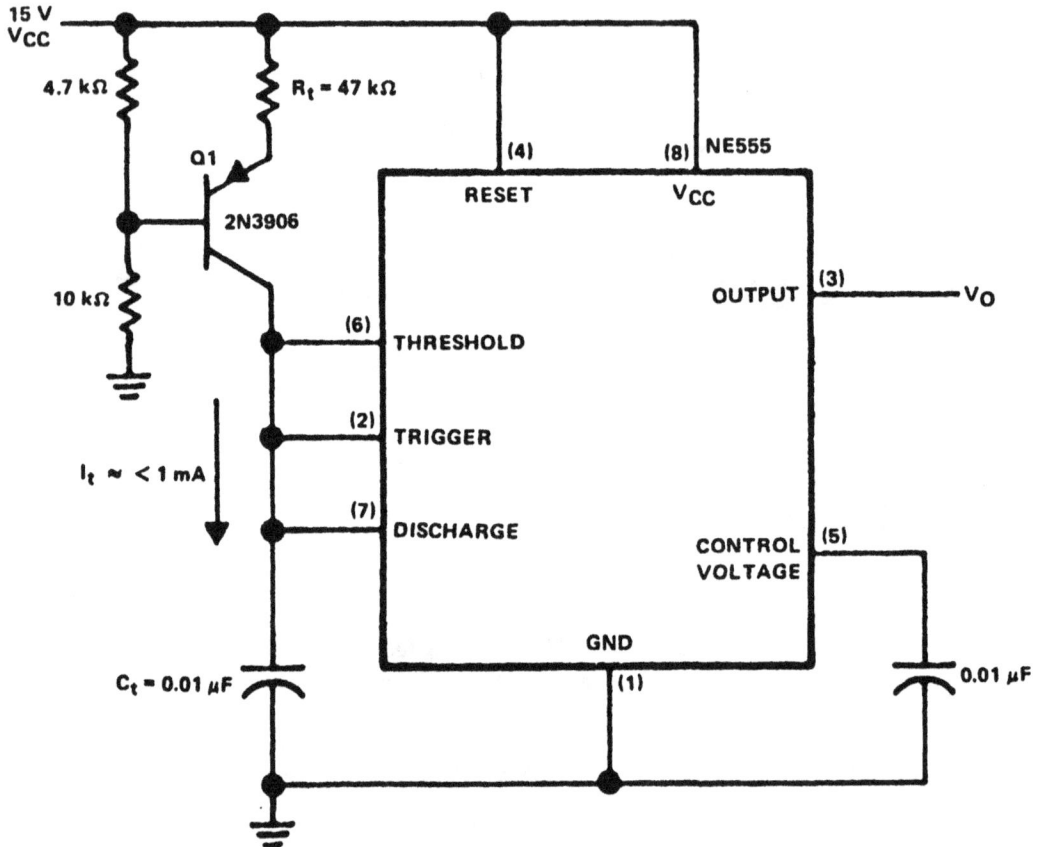

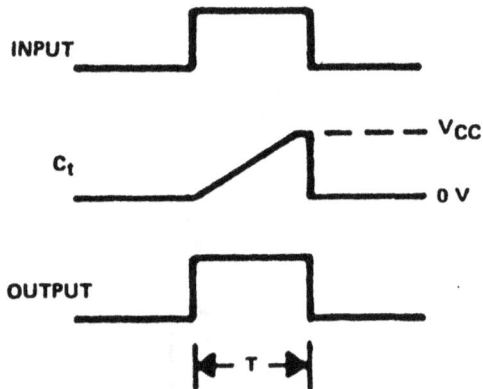

TEXAS INSTRUMENTS

Fig. 3-48

The linear charging ramp is most useful where linear control of voltage is required. Some possible applications are a long-period voltage-controlled timer, a voltage-to-pulse width converter, or a linear pulse width modulator. Q1 is the current source transistor, supplying constant current to the timing capacitor C_t. When the timer is triggered, the clamp on C_t is removed and C_t charges linearly toward V_{CC} by virtue of the constant current supplied by Q1. The threshold at pin 6 is $2/3\ V_{CC}$; here, it is termed V_C. When the voltage across C_t reaches V_C volts, the timing cycle ends. The timing expression for output pulse with T is:

In general, I_t should be at 1-mA, compatible with the NE555.

FUNCTION GENERATOR

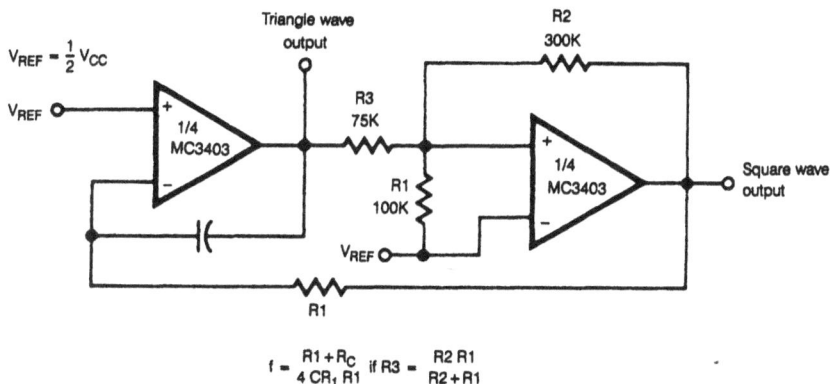

$$f = \frac{R1 + R_C}{4\ CR_1\ R1} \quad \text{if R3} = \frac{R2\ R1}{R2 + R1}$$

SIGNETICS

Fig. 3-49

105

ASTABLE MULTIVIBRATOR

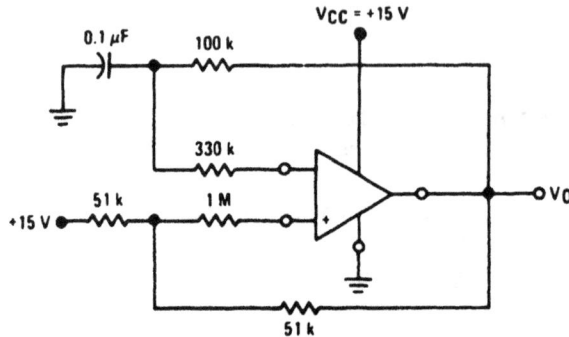

IC = MC3301

Fig. 3-50

WAVEFORM GENERATOR

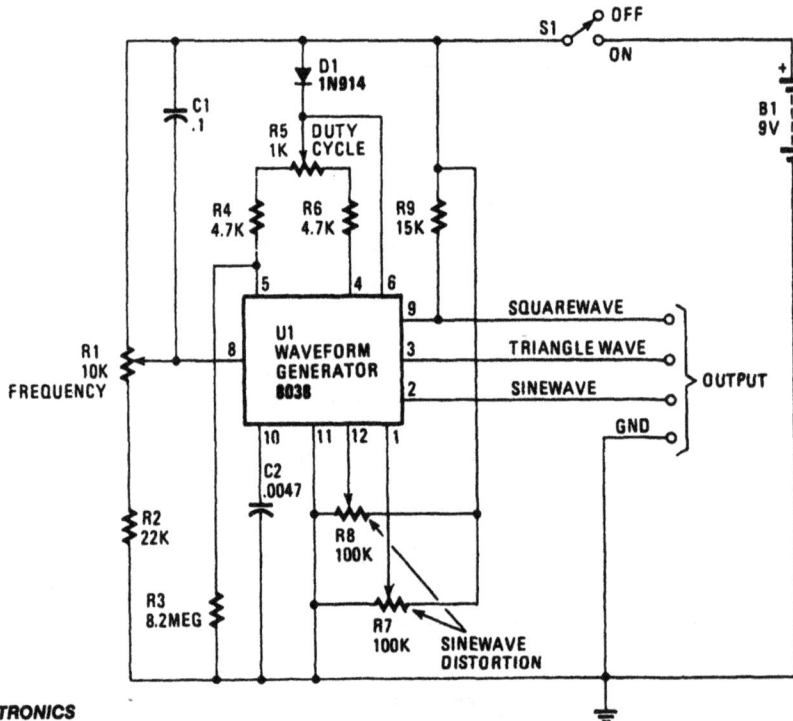

Fig. 3-51

The circuit is designed around the Intersil 8038CC. Frequency range is approximately 20 Hz to 20 kHz—a tuning range of 1000:1 with a single control. The output frequency depends on the value of C2 and on the setting of potentiometer R1. Other values of C2 change the frequency range. Increase the value of C2 to lower the frequency. The lowest possible frequency is around 0.001 Hz and the highest is around 300 kHz.

LAST-CYCLE COMPLETING GATED OSCILLATOR

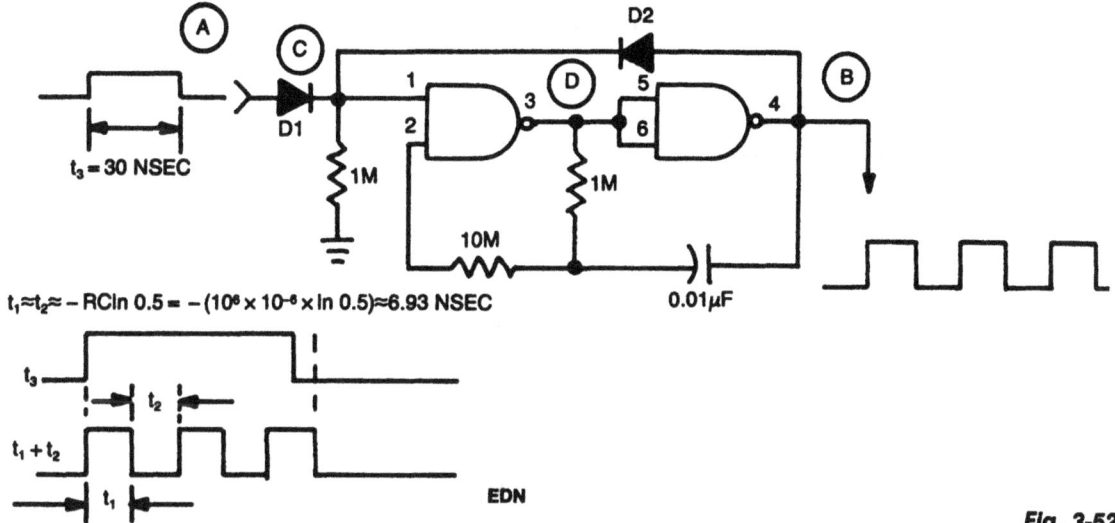

$$t_1 \approx t_2 \approx - RC\ln 0.5 = -(10^6 \times 10^{-6} \times \ln 0.5) \approx 6.93 \text{ NSEC}$$

EDN

Fig. 3-52

Regenerative feedback at C enables the oscillator to complete its timing cycle, rather than immediately shutting it off. The IC used was a CD4011AE, although an equivalent will work.

PRECISE TRIANGLE-/ SQUARE-WAVE GENERATOR

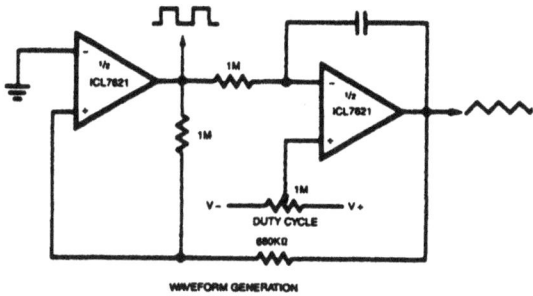

WAVEFORM GENERATION

INTERSIL

Fig. 3-53

Since the output range swings exactly from rail to rail, frequency and duty cycle are virtually independent of power supply variations.

WIDE-RANGE TRIANGLE-/ SQUARE-WAVE GENERATOR

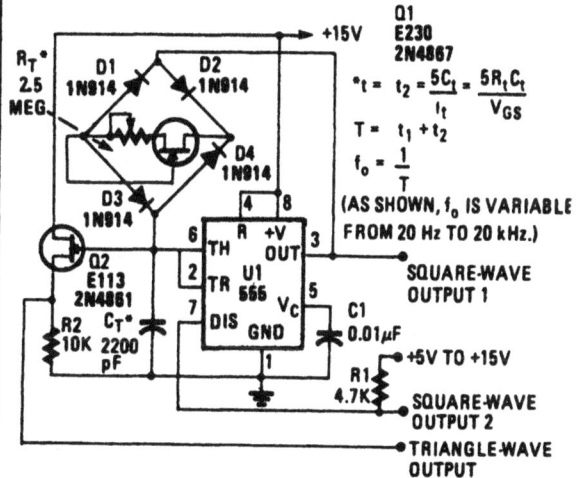

$$t = t_2 = \frac{5C_t}{I_t} = \frac{5R_tC_t}{V_{GS}}$$

$$T = t_1 + t_2$$

$$f_o = \frac{1}{T}$$

(AS SHOWN, f_o IS VARIABLE FROM 20 Hz TO 20 kHz.)

HANDS-ON ELECTRONICS

Fig. 3-54

107

WAVEFORM GENERATOR

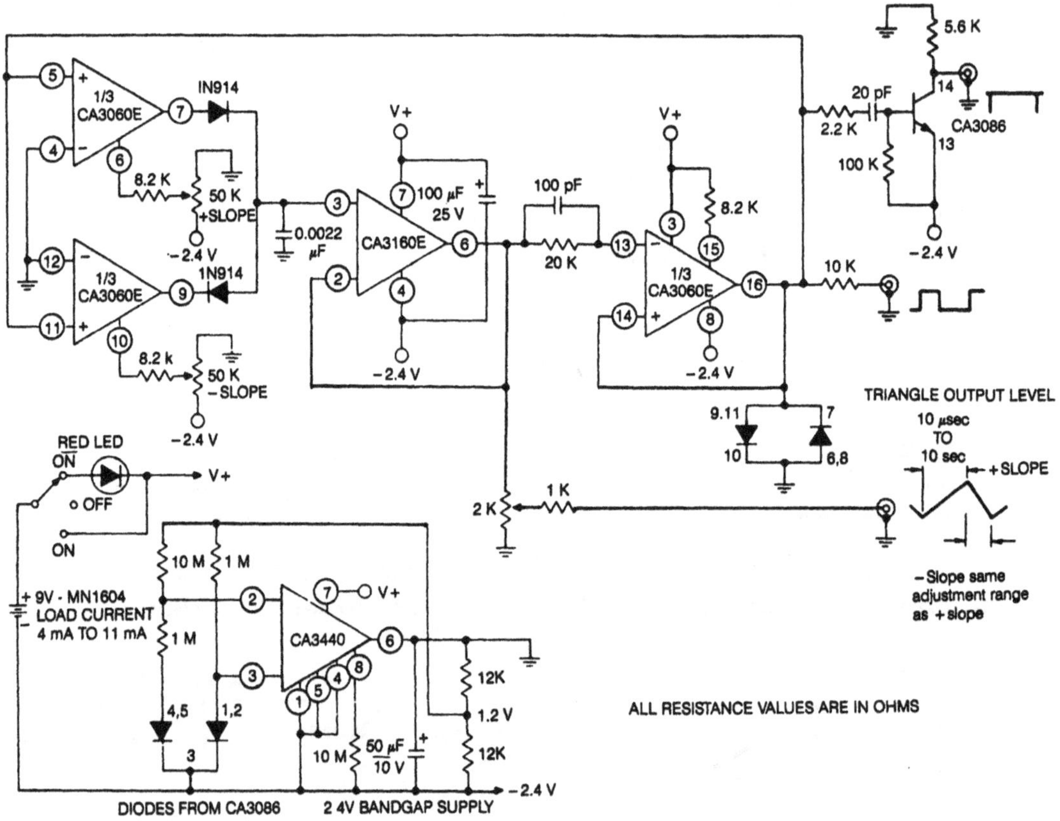

GENERAL ELECTRIC/RCA

Fig. 3-55

The circuit uses a CA3060 triple OTA (two units serve as switched current generators, controlled by a third amplifier). A CA3160 BiMOS op amp serves as a voltage follower to buffer the 0.0022-μF integrating capacitor. The circuit has an adjustment range of 1,000,000:1 and a timing range of 20 μs to 20 sec. The on-off switch actuates an LED that serves, as both a pilot light and a low-battery indicator. The LED extends battery life, because it drops battery voltage to the circuit by approximately 1.2 volts, thus reducing supply-current.

SINGLE-SUPPLY FUNCTION GENERATOR

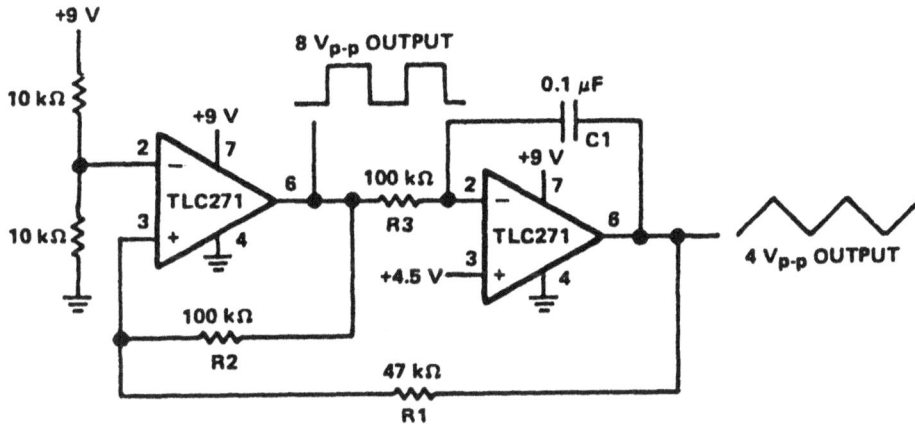

Fig. 3-56

This circuit has both square-wave and triangle-wave outputs. The left section is similar in function to a comparator circuit that uses positive feedback for hysteresis. The inverting input is biased at one-half the V_{CC} voltage by resistors R4 and R5. The output is fed back to the noninverting input of the first stage to control the frequency. The amplitude of the square wave is the output swing of the first stage, which is 8 V peak-to-peak. The second stage is basically an op-amp integrator. Resistor R3 is the input element and capacitor C1 is the feedback element. The ratio R1/R2 sets the amplitude of the triangle wave, as referenced to the square-wave output. For both waveforms, the frequency of oscillation can be determined by the equation:

$$f_o = \frac{1}{4R_3C_1} \frac{R_2}{R_1}$$

The output frequency is approximately 50 Hz with the given components.

PRECISE WAVE GENERATOR

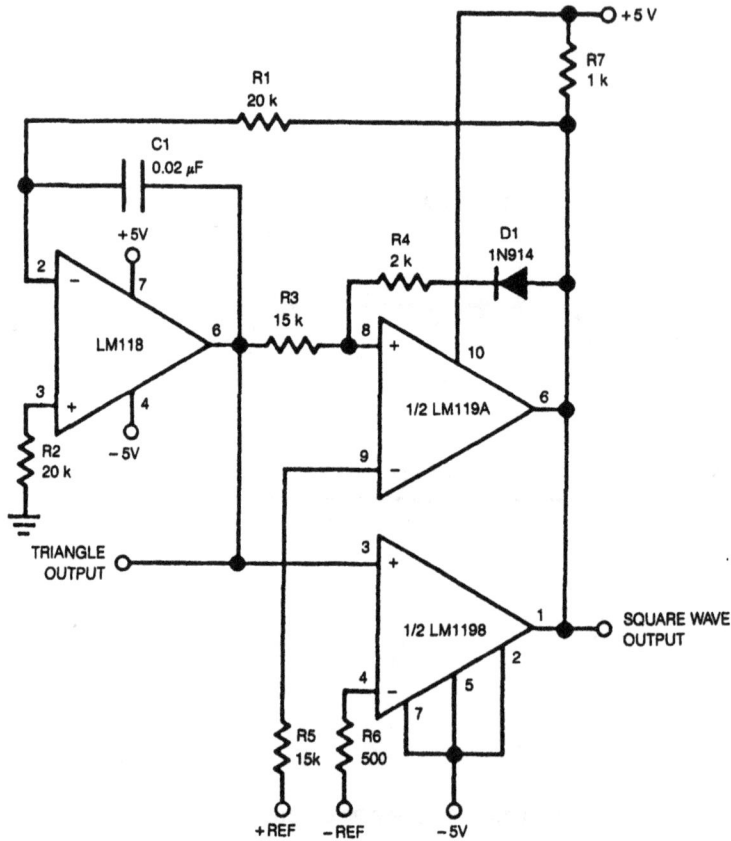

NATIONAL SEMICONDUCTOR

Fig. 3-57

The positive and negative peak amplitude is controllable to an accuracy of about ±0.01 V by a dc input. Also, the output frequency and symmetry are easily adjustable. The oscillator consists of an integrator and two comparators—one comparator sets the positive peak and the other sets the negative peak of the triangle wave. If R1 is replaced by a potentiometer, the frequency can be varied over at least a 10-to-1 range without affecting amplitude. Symmetry is also adjustable by connecting a 50-kΩ resistor from the inverting input of the LM118 to the arm of the 1-kΩ potentiometer. The ends of the potentiometer are connected across the supplies. Current for the resistor either adds or subtracts from the current through R1, which changes the ramp time.

FIXED-FREQUENCY VARIABLE DUTY-CYCLE OSCILLATOR

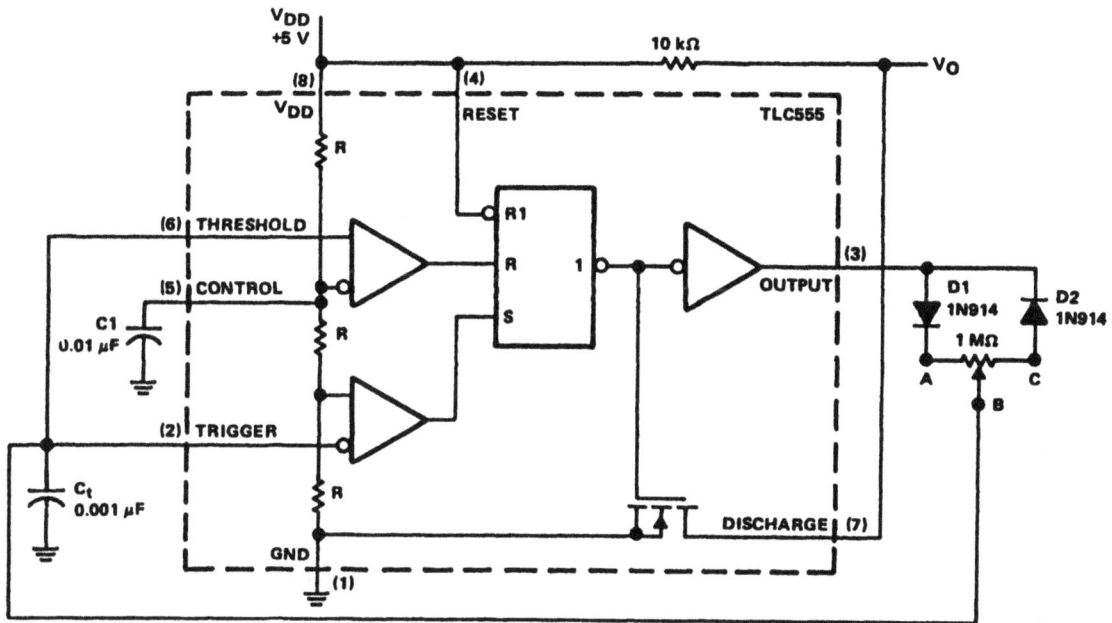

Fig. 3-58

In a basic astable timer, configuration timing periods t_1 and t_2 are not controlled independently. The lack of control makes it difficult to maintain a constant period, T, if either t_1 or t_2 is varied. In this circuit, charge R_{AB} and discharge R_{BC} resistances are determined by the position of common wiper arm B of the potentiometer. So, it is possible to adjust the duty-cycle by adjusting t_1 and t_2 proportionately, without changing period T.

At start-up, the voltage across C_t is less than the trigger level voltage ($1/2\ V_{DD}$), causing the timer to be triggered via pin 2. The output of the timer at pin 3 increases, turning off the discharge transistor at pin 7 and allowing C_t to charge through diode D1 and resistance R_{AB}. When capacitor C_t charges to upper threshold voltage $2/3\ V_{DD}$, the flip-flop is reset and the output at pin 3 decreases. Capacitor C_t then discharges through diode D2 and resistor R_{BC}. When the voltage at pin 2 reaches $1/3\ V_{DD}$, the lower threshold or trigger level, the timer triggers again and the cycle is repeated. In this circuit, the oscillator frequency remains fixed and the duty cycle is adjustable from less than 0.5% to greater than 99.5%.

WIDE-RANGE VARIABLE OSCILLATOR

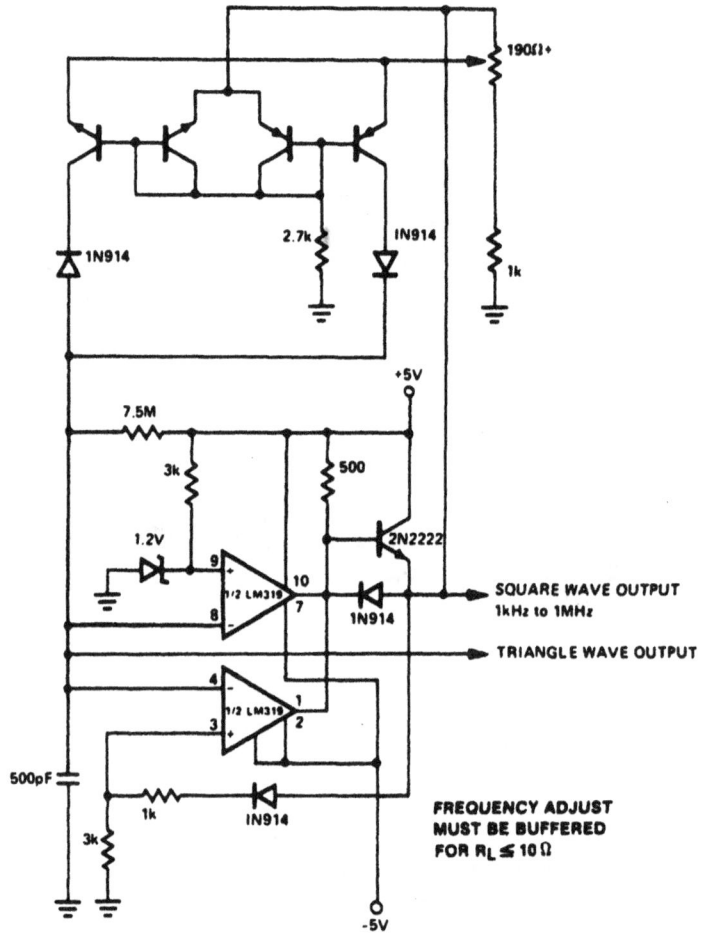

SQUARE WAVE OUTPUT
1kHz to 1MHz

TRIANGLE WAVE OUTPUT

FREQUENCY ADJUST
MUST BE BUFFERED
FOR $R_L \leq 10\,\Omega$

SIGNETICS

Fig. 3-59

SINGLE-CONTROL FOUR-DECADE VARIABLE OSCILLATOR

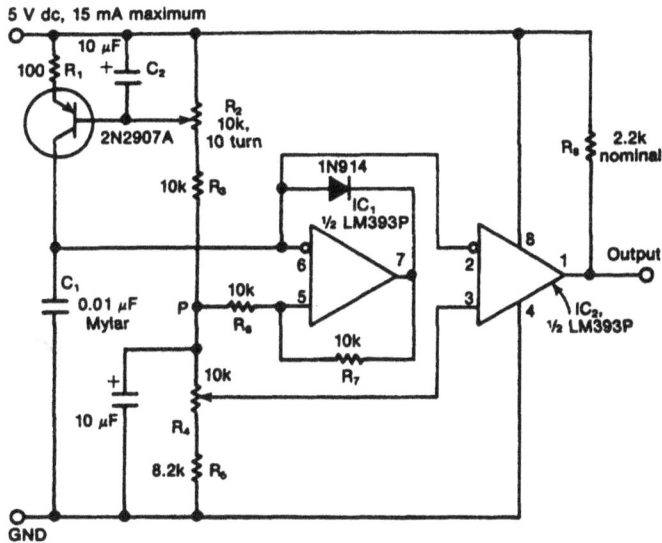

ELECTRONIC DESIGN

Fig. 3-60

The circuit consists of a variable current source that charges a capacitor, which is rapidly discharged by a Schmitt-trigger comparator. The sawtooth waveform thus produced is fed to another comparator, one with a variable switching level. The output from the second comparator is a pulse train with an independently adjustable frequency and duty cycle. The variable-frequency ramp generator consists of capacitor C1, which is charged by a variable and nonlinear current source. The latter comprises a 2N2907A pnp transistor, plus resistor R1 and the potentiometer R2. Capacitor C2 eliminates any ripple or noise at the base of the transistor that might cause frequency jitter at the output.

SIMPLE TRIANGLE SQUARE-WAVE OSCILLATOR

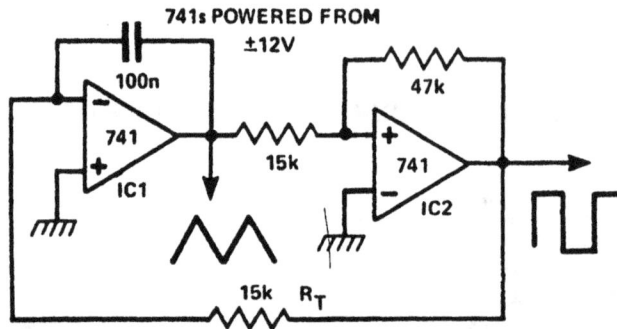

ELECTRONICS TODAY INTERNATIONAL   *Fig. 3-61*

This circuit generates simultaneously, a triangle and a square waveform. It is self-starting and has no latch-up problems. IC1 is an integrator with a slew rate which is determined by CT and RT, and IC2 is a Schmitt trigger. The output of IC1 ramps up and down between the hysteresis levels of the Schmitt, the output of which drives the integrator. By making RT variable, it is possible to alter the operating frequency over a 100-to-1 range. Three resistors, one capacitor, and a dual op amp is all that is needed to make a versatile triangle- and square-wave oscillator with a possible frequency range of 0.1 Hz to 100 kHz.

4

Miscellaneous Oscillators

The sources of the following circuits are contained in the Sources section, which begins on page 173. The figure number contained in the box of each circuit correlates to the source entry in the Sources section.

Tunable-Frequency Oscillator
Resistance-Controlled Digital Oscillator
Cassette Bias Oscillator
Oscillator Adjustable Over 10:1 Range
1-Second 1-kHz Oscillator
Sine-Wave Shaper
Low-Distortion Sine-Wave Oscillator
Tunable Single-Comparator Oscillator
Wide-Range Oscillator (Frequency Range of 5000:1)
RLC Oscillator
HC-Based Oscillator
Programmable-Frequency Sine-Wave Oscillator

Variable Wien-Bridge Oscillator
50% Duty-Cycle Oscillator
High-Frequency Oscillator
Low-Frequency Oscillator
Quadrature Oscillator
Wien-Bridge Oscillator
CMOS Oscillator
XOR-Gate Oscillator
SCR Relaxation Oscillator
CMOS Oscillator
5-V Oscillator
Low-Voltage Wien-Bridge Oscillator

TUNABLE-FREQUENCY OSCILLATOR

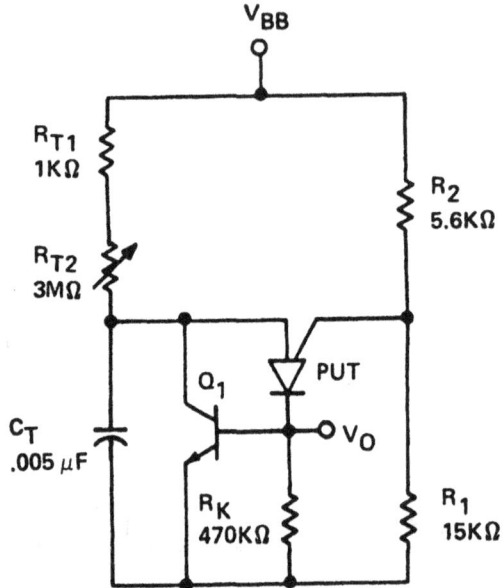

FREQUENCY RANGE
40 Hz to 65 kHz

OUTPUT PULSE

Rise time ~ 200 nsec.
Pulse width ~ 10 μsec.
Recovery time < 200 nsec.

Fig. 4-1A

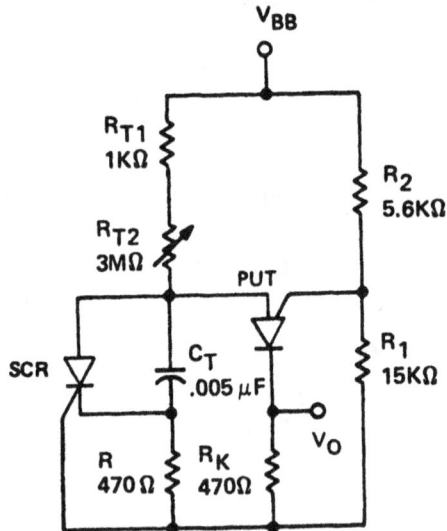

FREQUENCY RANGE
40 Hz to 40 kHz

OUTPUT PULSE

Width ~ 5 μsec.

UNITRODE CORPORATION

Fig. 4-1B

The variable oscillator circuit includes active elements for discharging the timing capacitor C_T (Fig. 4-1A). A second method is given as in Fig. 4-1B.

RESISTANCE-CONTROLLED DIGITAL OSCILLATOR

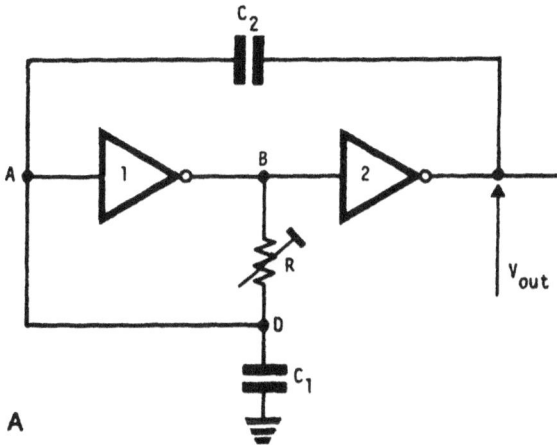

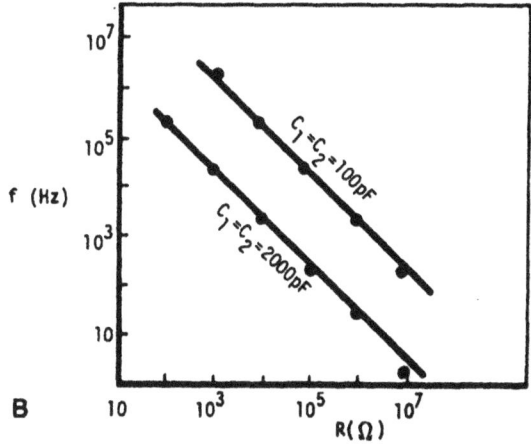

A

ELECTRONIC ENGINEERING

B

f (Hz)

$R(\Omega)$

$C_1=C_2=100pF$

$C_1=C_2=2000pF$

Fig. 4-2

This very simple, low-cost oscillator, is built with two CMOS buffer inverters, two capacitors and a variable resistance. The circuit can work with voltages ranging from 4 V up to 18 V. If C1 = C2, the frequency of oscillation, (ignoring the output and input impedance) is given by:

$$f = \frac{1}{4\pi\sqrt{2}RC}$$

The graph in Fig. B shows how the output frequency varies with resistance when C1 = C2 = 100 pF and C1 = C2 = 2000 pF.

CASSETTE BIAS OSCILLATOR

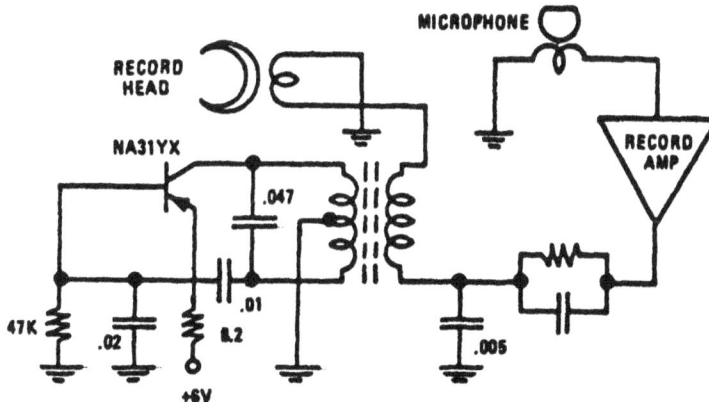

MICROPHONE

RECORD HEAD

NA31YX

RECORD AMP

.047

47K

.02

8.2

.01

.005

+6V

NATIONAL SEMICONDUCTOR

Fig. 4-3

OSCILLATOR ADJUSTABLE OVER 10:1 RANGE

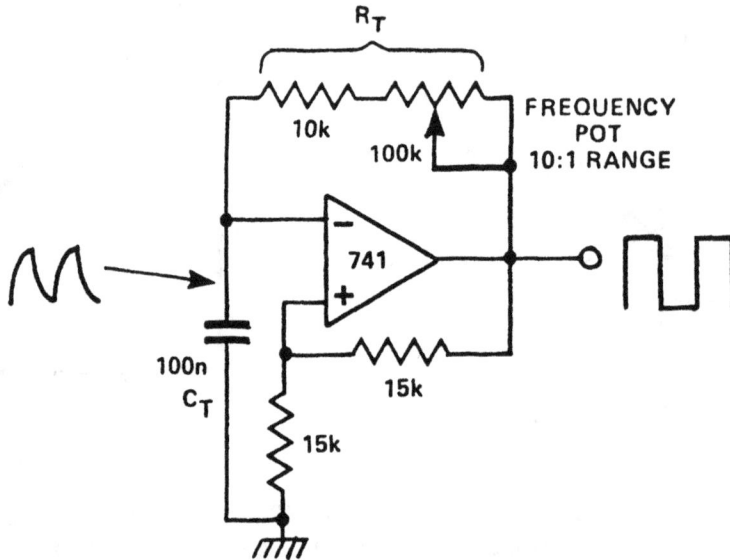

ELECTRONICS TODAY INTERNATIONAL

Fig. 4-4

In this circuit, two feedback paths are around an op amp. One is positive dc feedback, which forms a Schmitt trigger. The other is a CR timing network. Imagine that the output voltage is +10 V. The voltage at the noninverting terminal is +15 V. The voltage at the inverting terminal is a rising voltage with a time constant of $C_T R_T$. When this voltage exceeds +5 V, the op amp's output will go low and the Schmitt trigger action will make it snap into its negative state. Now the output is −10 V and the voltage at the inverting terminal falls with the time constant as before. By changing this time constant with a variable resistor, a variable frequency oscillation may be produced.

1-SECOND 1-kHz OSCILLATOR

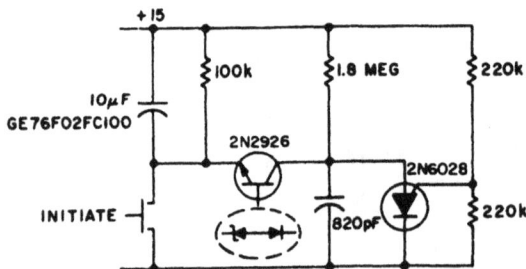

This circuit operates as an oscillator and a timer. The 2N6028 is normally on due to excess holding current through the 100-kΩ resistor. When the switch is momentarily closed, the 10-μF capacitor is charged to a full 15 volts and 2N2926 starts oscillating (1.8 MΩ and 820 pF). The circuit latches when 2N2926 zener breaks down again.

GENERAL ELECTRIC

Fig. 4-5

SINE-WAVE SHAPER

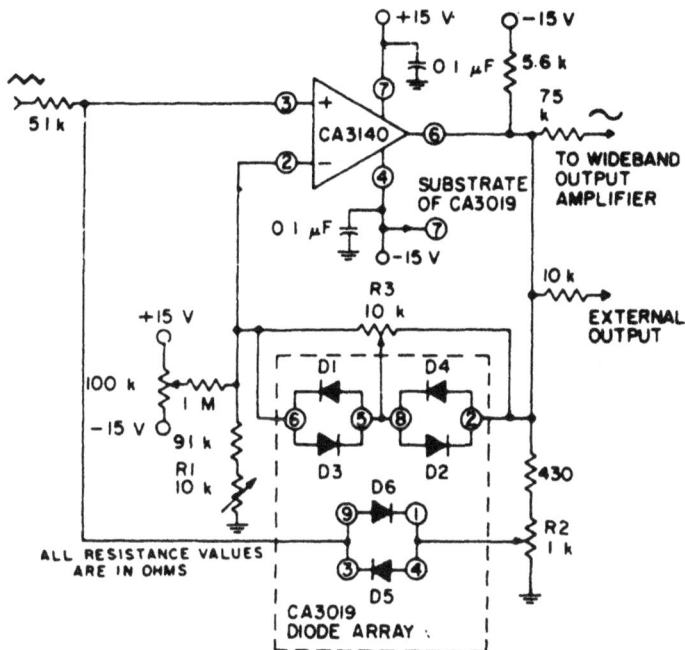

ALL RESISTANCE VALUES ARE IN OHMS

GENERAL ELECTRIC/RCA

Fig. 4-6

This circuit uses a CA3140 BiMOS op amp as voltage follower, together with diodes from a CA3019 array, to convert a triangular signal (such as obtained from a function generator) to a sine-wave output with typical THD less than 2%.

LOW-DISTORTION SINE-WAVE OSCILLATOR

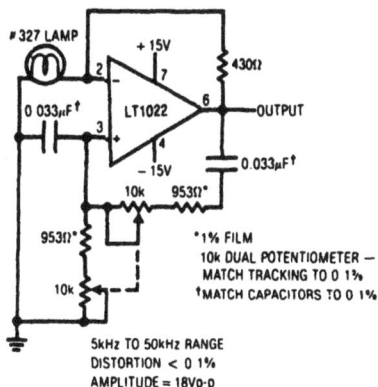

*1% FILM
10k DUAL POTENTIOMETER —
MATCH TRACKING TO 0 1%
†MATCH CAPACITORS TO 0 1%

5kHz TO 50kHz RANGE
DISTORTION < 0 1%
AMPLITUDE = 18Vp-p

Fig. 4-7

LINEAR TECHNOLOGY CORP.

TUNABLE SINGLE-COMPARATOR OSCILLATOR

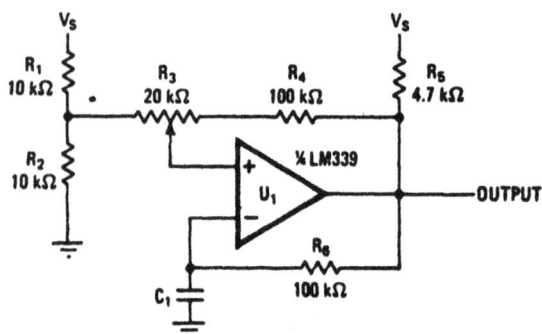

Varying the amount of this comparator circuit's hysteresis makes it possible to smoothly vary output frequencies in the 740-Hz to 2.7-kHz range. The amount of hysteresis together with time constant R6C1 determines how much time it takes for C1 to charge or discharge to the new threshold after the output voltage switches.

ELECTRONICS   *Fig. 4-8*

WIDE-RANGE OSCILLATOR (FREQUENCY RANGE OF 5000:1)

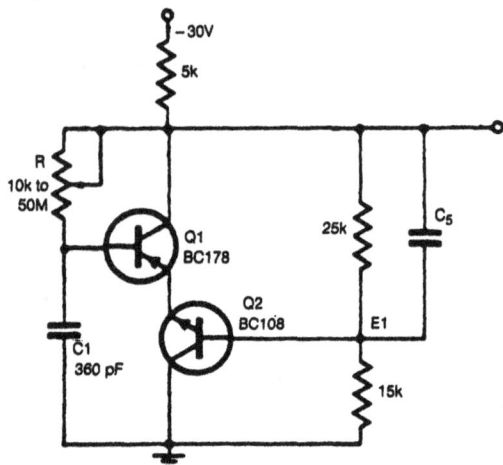

Timing resistor R can be adjusted to any value between 10 kΩ and 50 MΩ to obtain a frequency range from 400 kHz to 100 Hz. Returning the timing resistor to the collector of Q1 ensures that Q1 draws its base current only from the timing capacitor Ct. The timing capacitor recharges when the transistors are off, to a voltage equal to the base emitter voltage of Q2 plus the base emitter drops of Q1 and Q2. The transistors then start into conduction. Capacitor Cs is used to speed up the transition. A suitable value would be in the region of 100 pF.

ELECTRONICS TODAY INTERNATIONAL   *Fig. 4-9*

RLC OSCILLATOR

A positive transient, such as the power switch closing, charges C through L to a voltage above the supply voltage, if Q is sufficient. When the current reverses, the diode blocks and triggers the SCS. As the capacitor discharges, the anode gate approaches ground potential, depriving the anode of holding current. This turns off the SCS, and C charges to repeat the cycle.

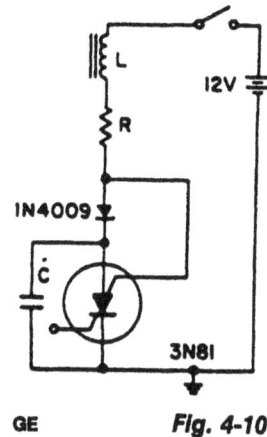

GE   *Fig. 4-10*

HC-BASED OSCILLATORS

Two inverters, one resistor, and one capacitor are all that is required to make a HC(T)-based oscillator that gives reliable operation up to about 10 MHz. The use of two HC inverters produces a fairly symmetrical rectangular output signal. In the same circuit, HCT inverters give a duty factor of about 25%, rather than about 50%, since the toggle point of an HC and an HCT inverter is $1/2 \, V_{CC}$, and slightly less than 2 V, respectively. If the oscillator is to operate above 10 MHz, the resistor is replaced with a small inductor, as shown in Fig. 4-11B.

The output frequency of the circuit in Fig. 4-11A is given as about 1/1.8rc, and can be made variable by connecting a 100-kΩ potentiometer in series with R. The solution adopted for the oscillator in Fig. 4-11B is even simpler: C is a 50-pF trimmer capacitor.

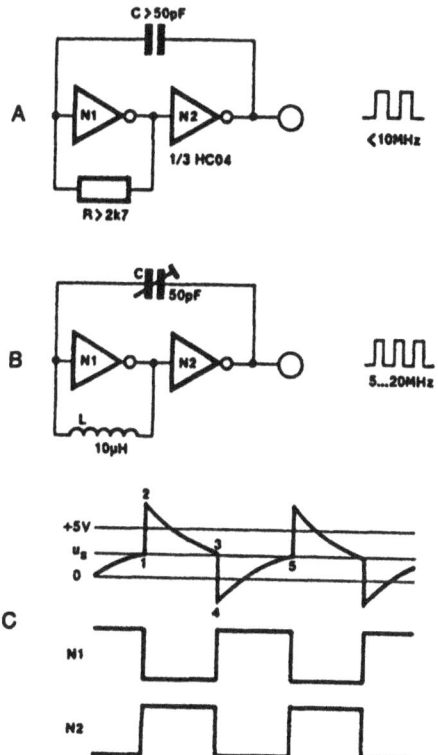

ELEKTOR ELECTRONICS   *Fig. 4-11*

121

PROGRAMMABLE-FREQUENCY SINE-WAVE OSCILLATOR

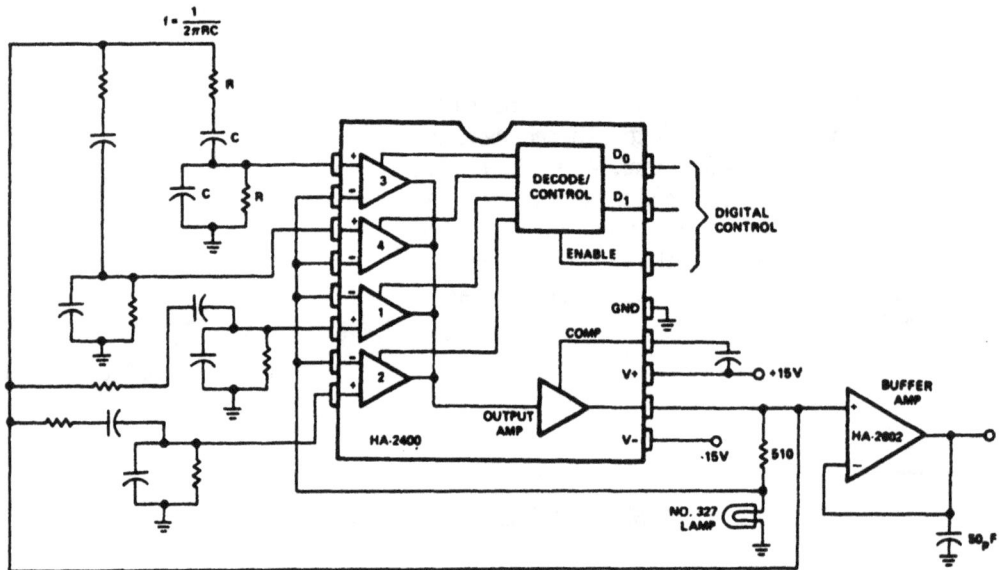

$$f = \frac{1}{2\pi RC}$$

HARRIS

Fig. 4-12

This Wien-bridge oscillator is very popular for signal generators, because it is easily turned over a wide frequency range, and has a very low distortion sine-wave output. The frequency determining networks can be designed from about 10 Hz to greater than 1 MHz; the output level is about 6.0-V rms. By substituting a programmable attenuator for the buffer amplifier, a very versatile sine-wave source for automatic testing, etc. can be constructed.

VARIABLE WIEN-BRIDGE OSCILLATOR

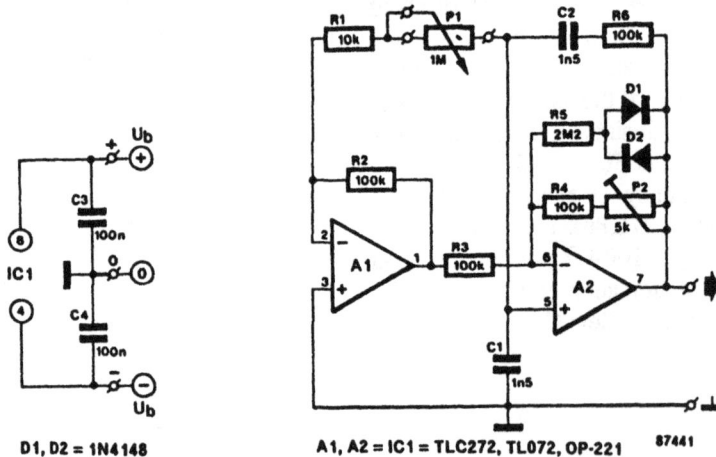

D1, D2 = 1N4148

A1, A2 = IC1 = TLC272, TL072, OP-221

ELEKTOR ELECTRONICS

Fig. 4-13

VARIABLE WIEN-BRIDGE OSCILLATOR (continued)

A Wien-bridge oscillator can be made variable by using two frequency-determining parts that are varied simultaneously at high tracking accuracy. High-quality tracking potentiometers or variable capacitors are, however, expensive and difficult to obtain. To avoid having to use such a component, this oscillator was designed to operate with a single potentiometer. The output frequency, f_o, is calculated from:

$$f_o = \frac{1}{(2nRCva)}$$

where:

$$R = R_2 = R_3 = R_4 = R_6, \ C = C_1 = C_2, \text{ and } a = (P_1 + R_1)R$$

With preset P2 you can adjust the overall amplification so that the output signal has a reasonably stable amplitude, 3.5 V_{pp} max., over the entire frequency range. The stated components allow the frequency to be adjusted between 350 Hz and 3.5 kHz.

50% DUTY-CYCLE OSCILLATOR

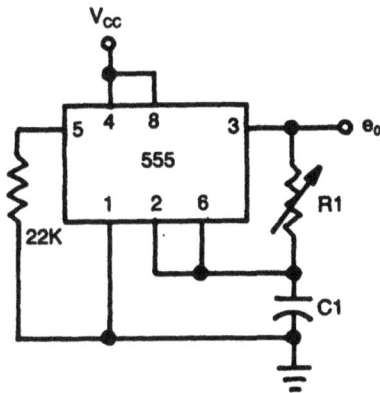

EDN   **Fig. 4-14**

Frequency of oscillation depends on the R1/C1 time constant and allows frequency adjustment by varying R1.

HIGH-FREQUENCY OSCILLATOR

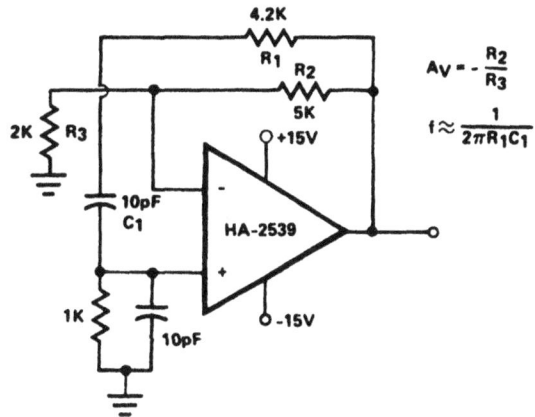

$$A_V = -\frac{R_2}{R_3}$$

$$f \approx \frac{1}{2\pi R_1 C_1}$$

HARRIS   **Fig. 4-15**

Intended primarily as a building block for a QRP transmitter, this 20-MHz oscillator delivered a clean 6-V, pk-pk signal into a 100-Ω load.

LOW-FREQUENCY OSCILLATOR

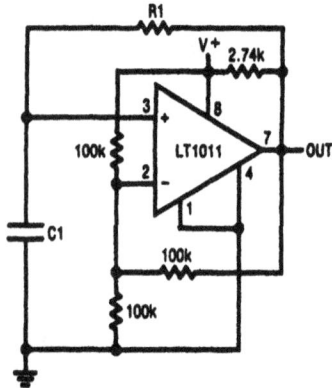

This simple RC oscillator uses a medium-speed comparator with hysteresis and feedback through R1 and C1 as timing elements. The frequency of oscillation is, at least theoretically, independent from the power supply voltage. If the comparator swings to the supply rails, if the pull-up resistor is much smaller than the resistor R_h, and if the propagation delay is negligible compared to the RC time constant, the oscillation frequency is:

$$f_{OSC} = \frac{0.72}{R_1 C_1}$$

LINEAR TECHNOLOGY CORP.   *Fig. 4-16*

QUADRATURE OSCILLATOR

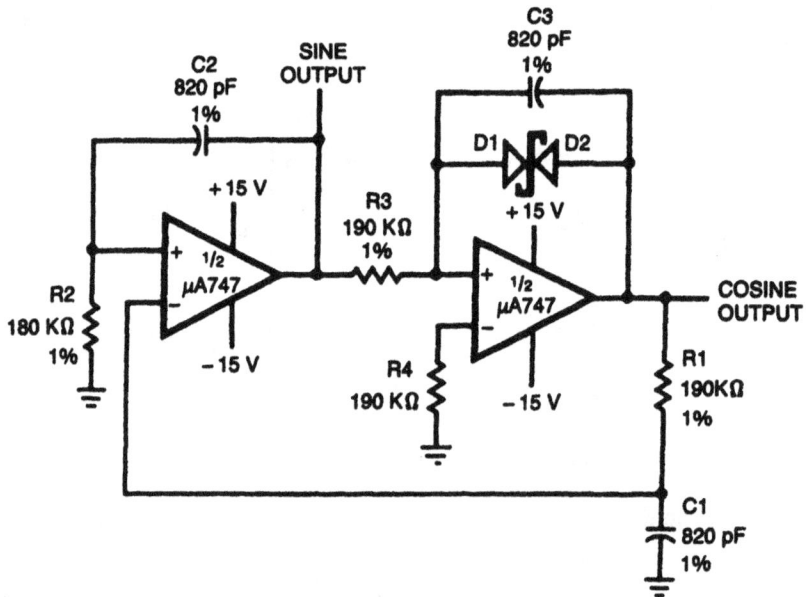

FAIRCHILD CAMERA AND INSTRUMENT   *Fig. 4-17*

WIEN-BRIDGE OSCILLATOR

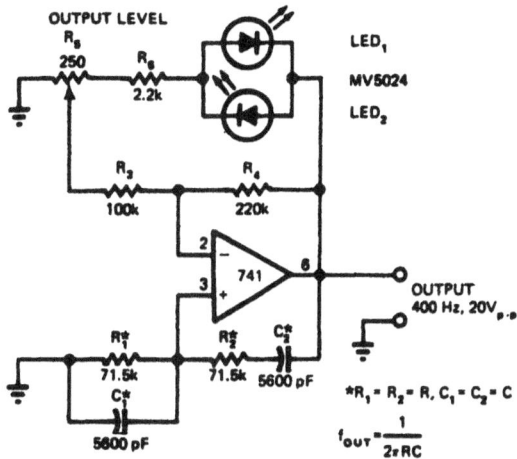

$$*R_1 = R_2 = R, C_1 = C_2 = C$$

$$f_{OUT} = \frac{1}{2\pi RC}$$

Fig. 4-18

LEDs function as both pilot lamps and as an AGC (automatic gain control) in this unconventional amplitude-stabilized oscillator.

CMOS OSCILLATOR

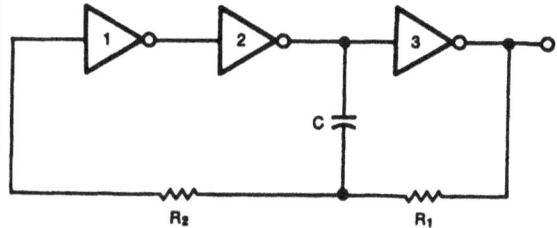

Fig. 4-19

This circuit is guaranteed to oscillate at a frequency of about $2.2/(R_1 \times C)$ if R_2 is greater than R_1. You can reduce the number of gates further if you replace gates 1 and 2 with a noninverting gate.

XOR-GATE OSCILLATOR

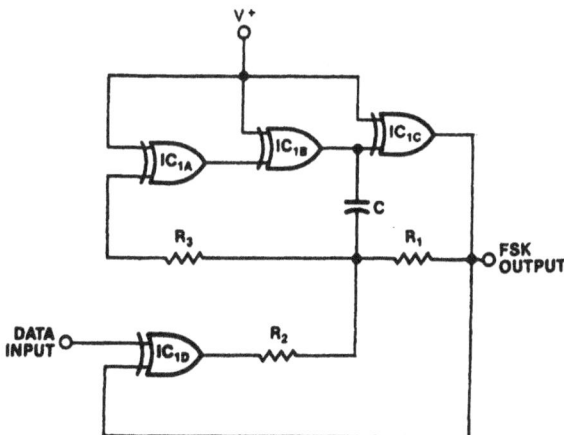

NOTES:
$IC_1 = CD4070B$
$R_1 \geq 15k$
$R_2, R_3 \geq 3R_1$
$f_o = \frac{1}{2R_1 C}$
$V^+ = 5$ TO 15V

Fig. 4-20

An exclusive-OR gate, IC1D, turns a simple CMOS oscillator into an FSK generator. When the data input increases, IC1D inverts, and negative feedback through R2 lowers the circuit's output frequency. A low input results in positive feedback and a higher output frequency. R1 and C set the oscillator's frequency range, and R2 determines the circuit's frequency shift. To ensure frequency stability, make R3 much greater than R1 and use a high-quality feedback capacitor. The three gates constituting the oscillator itself need not be exclusive-OR types; use any CMOS inverter.

SCR RELAXATION OSCILLATOR

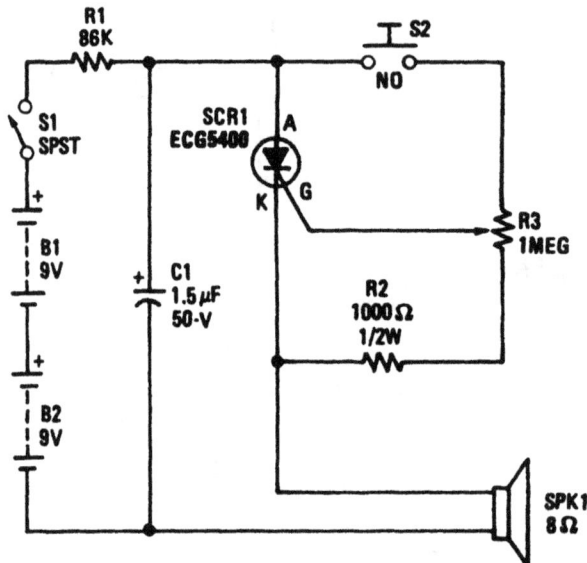

Fig. 4-21

CMOS OSCILLATOR

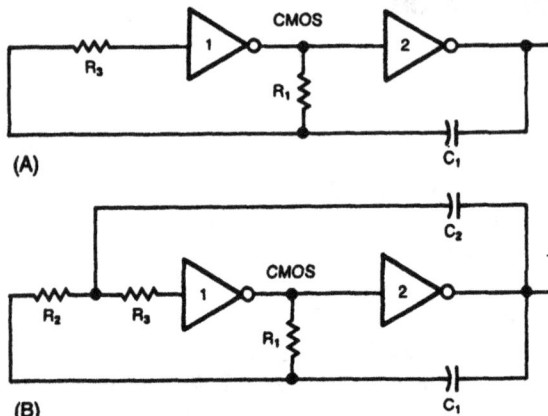

(A)

(B)

EDN

Fig. 4-22

cuit does work, oscillation occurs usually because both gates are in the package and, therefore, have logic thresholds only a few millivolts apart.

The circuit in Fig. 4-22B resolves both problems by adding a resistor and a capacitor. The R2/C2 network provides hysteresis, thus delaying the onset of gate 1's transition until C1 has enough voltage to move gate 1 securely through its transition region. When gate 1 is finally in its transition region, C2 provides positive feedback, thus rapidly moving gate 1 out of its transition region.

The equations for the oscillator in Fig. 4-22B are:

$$R_2 = 10R_1$$
$$R_3 = 10R_2$$
$$C_1 = 100C_2$$

$$f \cong \frac{1}{1.2R_1C_1}$$

The common clock oscillator in Fig. 4-22A has two small problems: It might not, in fact, oscillate if the transition regions of its two gates differ. If it does oscillate, it might sometimes oscillate at a slightly lower frequency than its equation predicts because of the finite gain of the first gate. If the cir-

5-V OSCILLATOR

Consistently self-starting and yet capable of operating from over 1 Hz to 10 MHz, this low-cost oscillator requires only five components. Calculate the period of oscillation by using this relationship: $P = 5 \times 10^3\ C$ sec when $C = C_1 = C_2$. By changing the ratio of C1 to C2, the duty cycle can be as low as 20%.

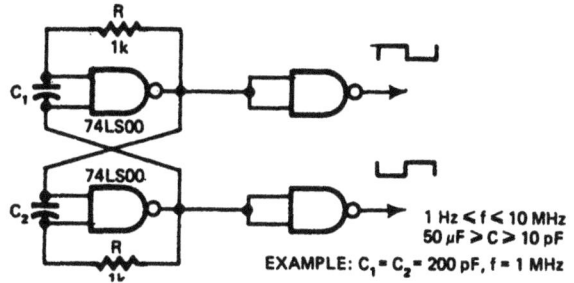

1 Hz $<$ f $<$ 10 MHz
50 μF $>$ C $>$ 10 pF

EXAMPLE: $C_1 = C_2 = 200$ pF, f = 1 MHz

EDN

Fig. 4-23

LOW-VOLTAGE WIEN-BRIDGE OSCILLATOR

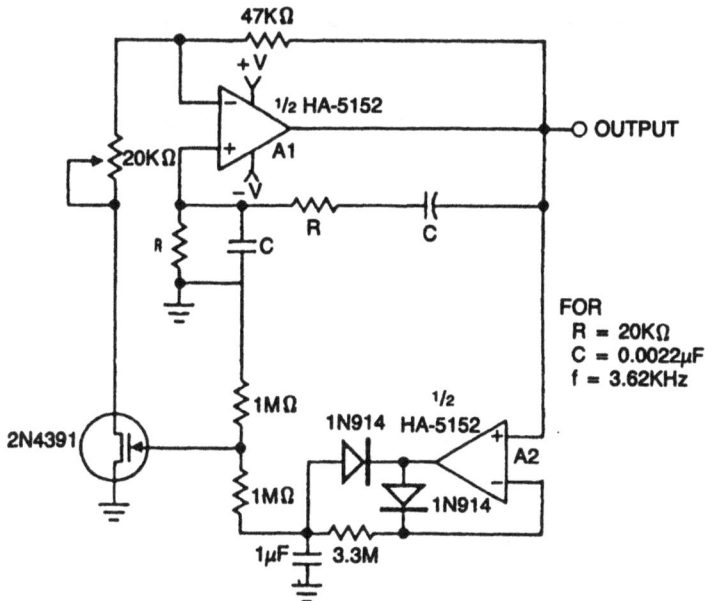

FOR
R = 20KΩ
C = 0.0022μF
f = 3.62KHz

HARRIS

Fig. 4-24

This circuit utilizes an HA-5152 dual op amp and FET to produce a low-voltage, low-power, Wein-bridge sine-wave oscillator. Resistors R and capacitors C control the frequency of oscillation; the FET, used as a voltage-controlled resistor, maintains the gain of A1 exactly 3 dB to sustain oscillation. The 20-kΩ pot can be used to vary the signal amplitude. The HA-5152 has the capability to operate from ±1.5-V supplies. This circuit will produce a low-distortion sine-wave output while drawing only 400 μA of supply current.

5

Multivibrators and Square-Wave Oscillators

The sources of the following circuits are contained in the Sources section, which begins on page 173. The figure number contained in the box of each circuit correlates to the source entry in the Sources section.

TTL OSCILLATOR

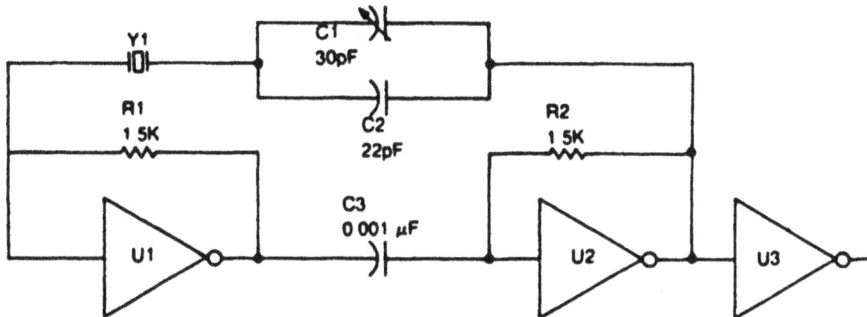

TAB BOOKS

Fig. 5-1

TTL inverter stages, U1 and U2, are cross-connected with a crystal Y1. A resistor in each stage biases the normally digital gates into a region where they operate as amplifiers. Inverter stage U3 is used as a buffer.

SQUARE-WAVE OSCILLATOR

Oscillator Frequency for Various Capacitor Values

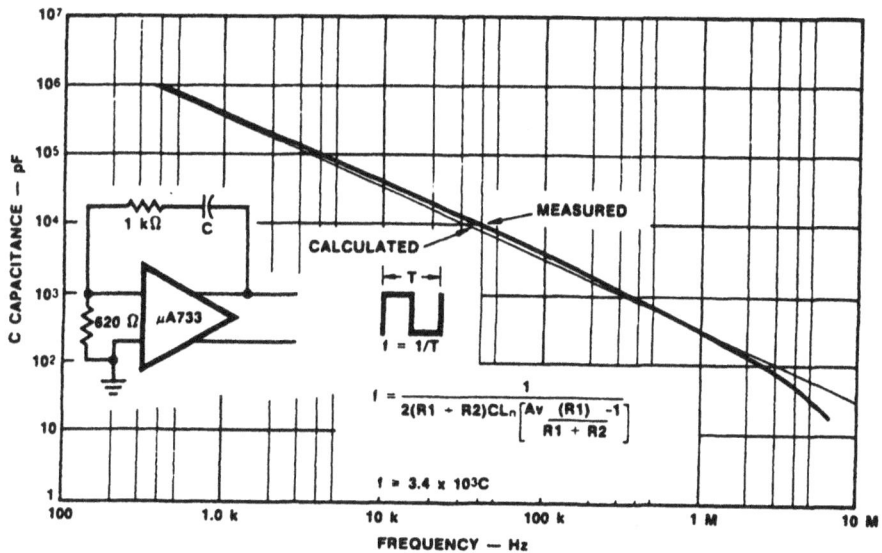

$$f = \frac{1}{2(R1 + R2)CL_n\left[Av\frac{(R1)}{R1 + R2} -1\right]}$$

$$f = 3.4 \times 10^3 C$$

FAIRCHILD CAMERA

Fig. 5-2

OSCILLATOR/CLOCK GENERATOR

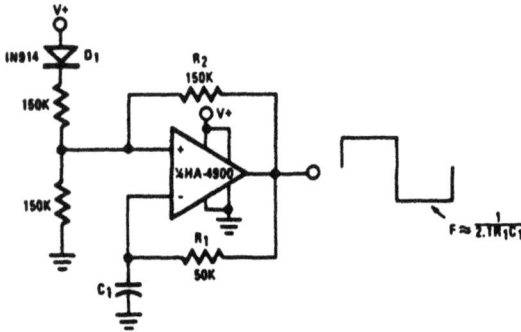

This self-starting fixed frequency circuit gives excellent frequency stability. R1 and C1 compose the frequency determining network while R2 provides the regenerative feedback. Diode D1 enhances the stability by compensating for the difference between V_{OH} and V_{Supply}. In applications where a precision clock generator up to 100 kHz is required, such as in automatic test equipment, C1 can be replaced by a crystal.

HARRIS SEMICONDUCTOR   *Fig. 5-3*

CMOS OSCILLATOR

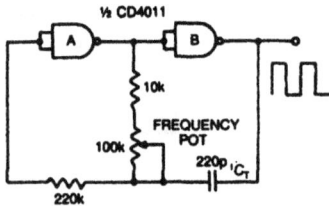

Varying the 100-kΩ pot changes the discharge rate of C_T and hence the frequency. A square-wave output is generated. The maximum frequency using CMOS is limited to 2 MHz.

ELECTRONICS TODAY INTERNATIONAL   *Fig. 5-4*

FREE-RUNNING SQUARE-WAVE OSCILLATOR

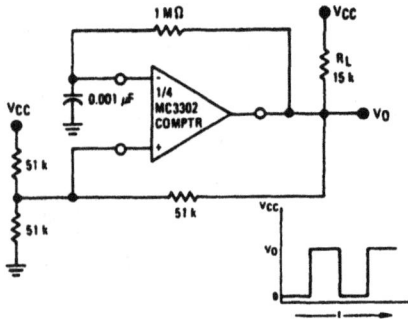

MOTOROLA   *Fig. 5-5*

PRECISION SQUARER

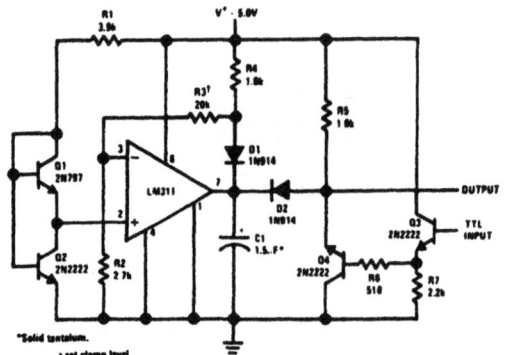

NATIONAL SEMICONDUCTOR   *Fig. 5-6*

SQUARE-WAVE OSCILLATOR

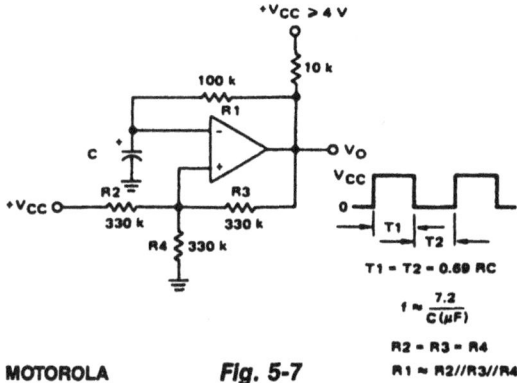

$$T1 = T2 = 0.69\ RC$$

$$f \approx \frac{7.2}{C(\mu F)}$$

R2 = R3 = R4

R1 ≈ R2//R3//R4

MOTOROLA   *Fig. 5-7*

0.5 Hz SQUARE-WAVE OSCILLATOR

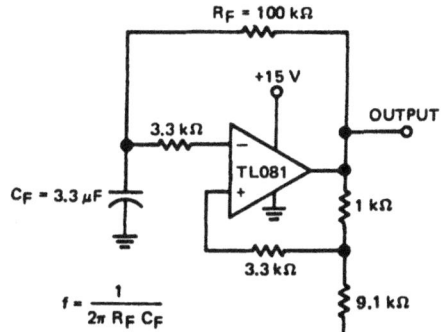

$$f = \frac{1}{2\pi\ R_F\ C_F}$$

TEXAS INSTRUMENTS   *Fig. 5-8*

SIMPLE TRIANGLE-/SQUARE-WAVE OSCILLATOR

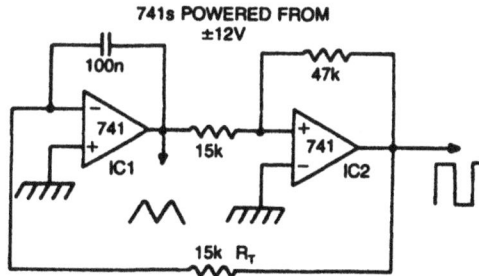

ELECTRONICS TODAY INTERNATIONAL   *Fig. 5-9*

By making R_T variable it is possible to alter the operating frequency over a 100-to-1 range. Versatile triangle-/square-wave oscillator has a possible frequency range of 0.1 Hz to 100 kHz.

SQUARE-WAVE OSCILLATOR

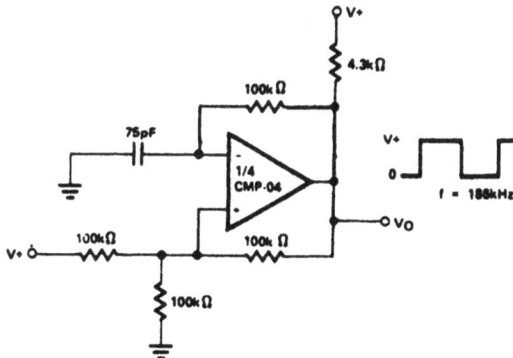

PRECISION MONOLITHICS

Fig. 5-10

RC OSCILLATOR

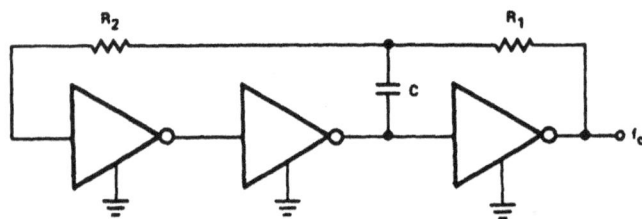

$$f_o \approx \frac{1}{2\,C[0.41\,R_P + 0.70\,R_1]} \quad , \quad R_P = \frac{R_1\,R_2}{R_1 + R_2}$$

a. If $R_1 = R_2 = R_1$, $f \approx 0.55/RC$

b. If $R_2 >> R_1$, $f \approx 0.45/R_1C$

c. If $R_2 << R_1$, $f \approx 0.72/R_1C$

a. $f = 120$ kHz, $C = 420$ pF
   $R_1 = R_2 \approx 10.9$ k Ω

b. $f = 120$ kHz, $C = 420$ pF, $R_2 = 50$ k Ω
   $R_1 = 8.93$ k Ω

c. $f = 120$ kHz, $C = 220$ pF, $R_2 = 5$ k Ω
   $R_1 = 27.3$ k Ω

Gates are 74C04

TELEDYNE

Fig. 5-11

1-kHz SQUARE-WAVE OSCILLATOR

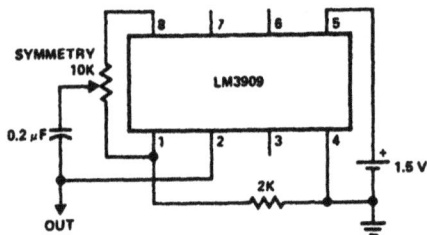

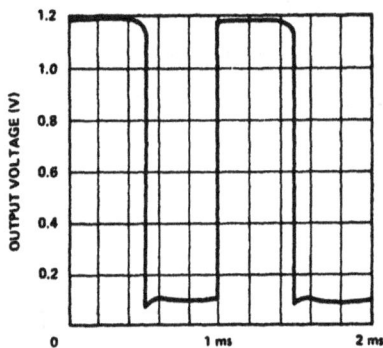

Note: Output Voltage Through a 10K Load to Ground

SILICONIX

Fig. 5-12

ADJUSTABLE TTL CLOCK (MAINTAINS 50% DUTY CYCLE)

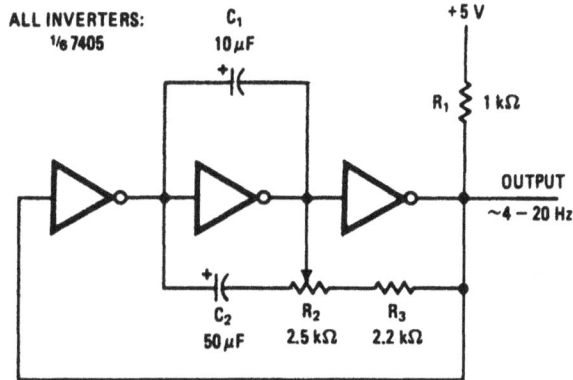

ELECTRONICS

Fig. 5-13

Symmetry of the square-wave output is maintained by connecting the right side of R2 through resistor R3 to the output of the third amplifier stage. This changes the charging current to the capacitors in proportion to the setting of frequency-adjusting potentiometer R2. Thus, a duty cycle of 50% is constant over the entire range of oscillation. The lower frequency limit is set by capacitor C2. With the components shown, the frequency of oscillation can be varied by R2 from about 4 to 20 Hz. Other frequency ranges can be obtained by changing the values of C1 and R3, which control the upper limit of oscillation, or C2, which limits the low-frequency end.

SQUARE-WAVE OSCILLATOR

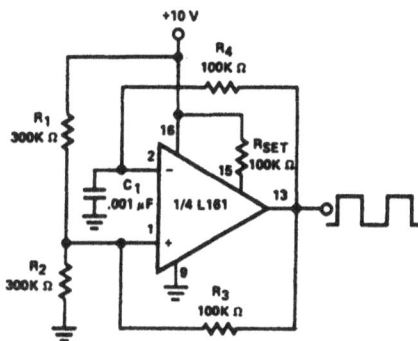

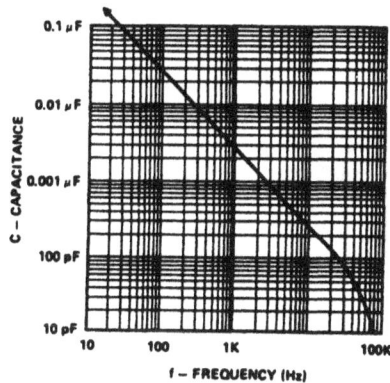

Frequency vs the Value of C_1
for the Squarewave Oscillator

SILICONIX

Fig. 5-14

This generator is operable to over 100 kHz. The low frequency limit is determined by C1. Frequency is constant for supply voltages down to +5 V.

DUAL ASTABLE MULTIVIBRATOR

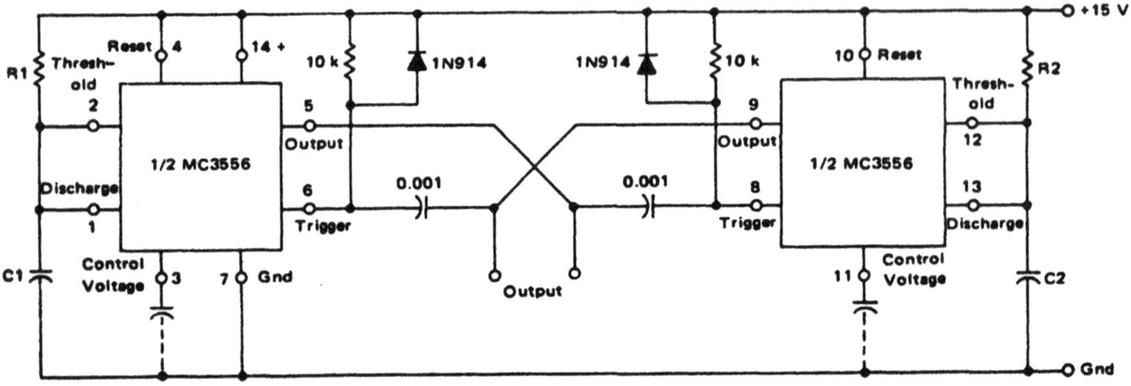

$$f = \frac{0.91}{(R1+R2)\,C} \quad \text{for } C1 = C2 \qquad \text{Duty Cycle } \frac{R2}{R1+R2}$$

MOTOROLA

Fig. 5-15

This dual astable multivibrator provides versatility not available with single-timer circuits. The duty cycle can be adjusted from 5% to 95%. The two outputs provide two phase clock signals often required in digital systems. It can also be inhibited by use of either reset terminal.

UJT MONOSTABLE

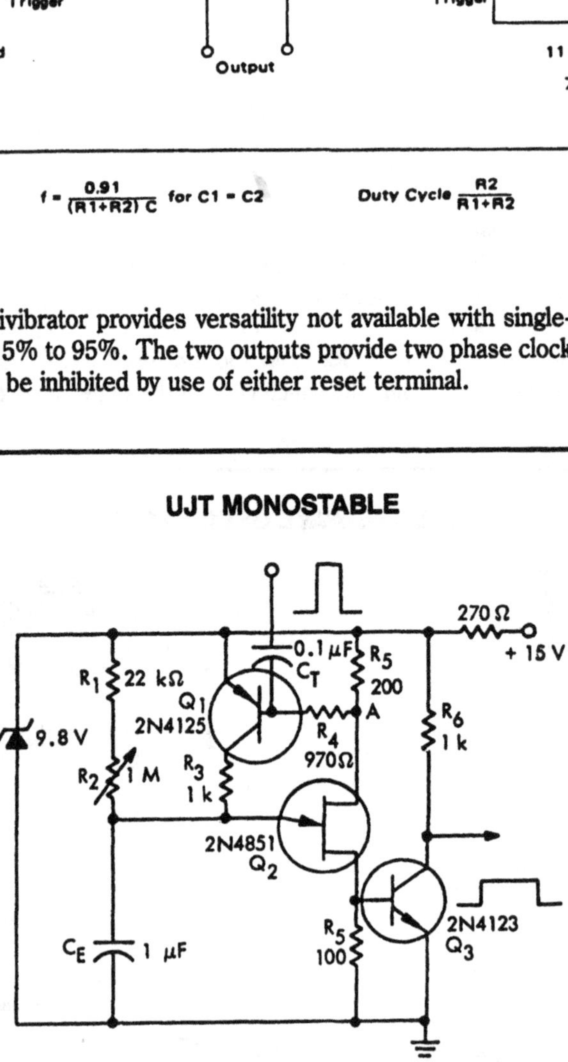

MOTOROLA

Fig. 5-16

MONOSTABLE MULTIVIBRATOR WITH INPUT LOCK-OUT

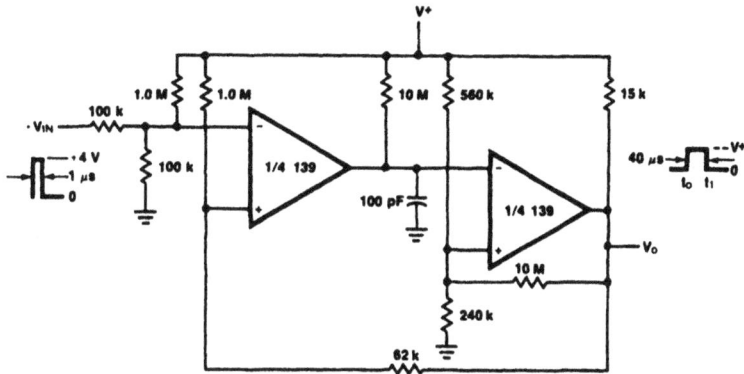

Fig. 5-17

TTL MONOSTABLE

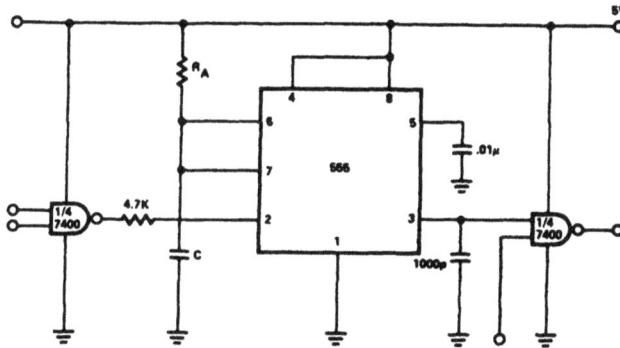

Fig. 5-18

MONOSTABLE

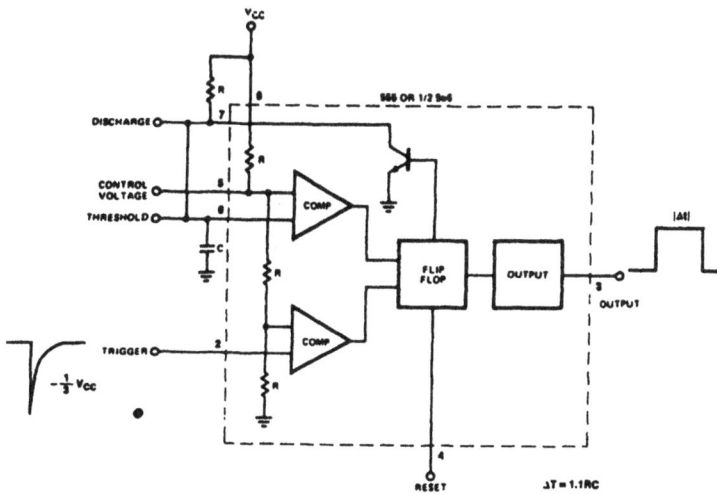

Fig. 5-19

ONE-SHOT MULTIVIBRATOR

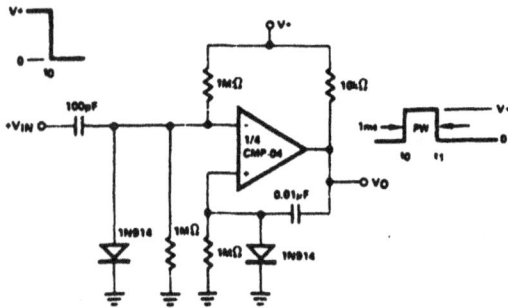

PRECISION MONOLITHICS   *Fig. 5-20*

BISTABLE MULTIVIBRATOR

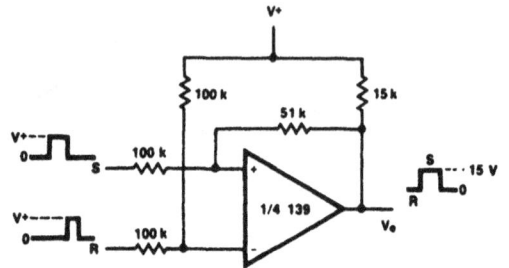

FAIRCHILD CAMERA   *Fig. 5-22*

MONOSTABLE MULTIVIBRATOR

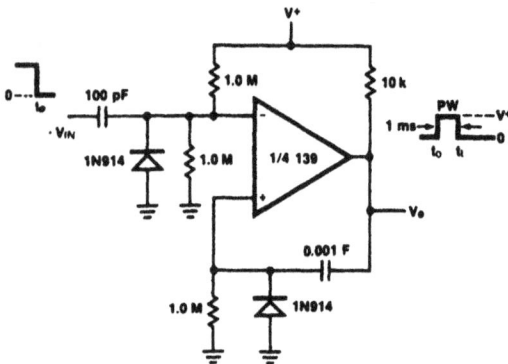

FAIRCHILD CAMERA   *Fig. 5-21*

100-kHz FREE-RUNNING MULTIVIBRATOR

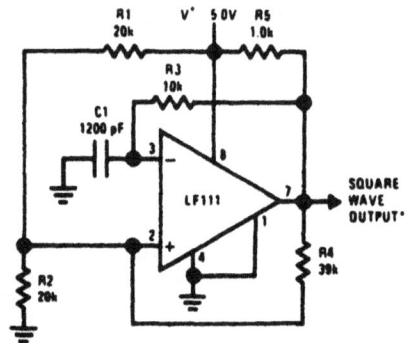

*TTL or DTL fanout of two.

NATIONAL SEMICONDUCTOR   *Fig. 5-23*

SQUARE-WAVE PULSE EXTRACTOR

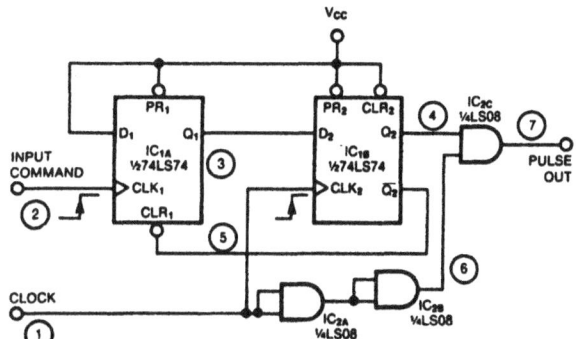

This circuit traps a single positive pulse from a square-wave train. Following the rising edge of an input command, the pulse-out signal emits a replica of one positive pulse of the clock signal simultaneous with the clock signal's next rising edge. The input command signal sets the Q1 output of flip-flop IC1A. Consequently, the next rising edge of the clock signal sets the Q2 output of IC1B, which allows AND gate IC2C to pass the clock signal's next positive pulse. AND gates IC2A and IC2B prevent the generation of brief output glitches by delaying the clock signal by t_D seconds (two propagation delays).

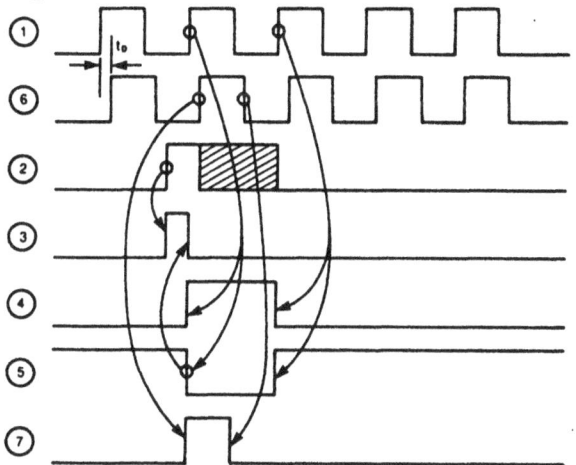

EDN

Fig. 5-24

NEARLY 50% DUTY-CYCLE MULTIVIBRATOR

Three factors contribute to the output symmetry. The capacitor charges and discharges through the same external resistor. An internal resistive divider sets accurate switching thresholds within the chip, the bipolar types use dividers, as well. Most importantly, IC1's CMOS output stage switches fully between ground and V_{CC}, avoiding the errors from asymmetry that are often found in a TTL timer's output. The IC's internal switching threshold tolerances can cause a deviation of several percent from the desired 50% duty cycle. To meet a tighter specification, you might have to select from a group of ICs.

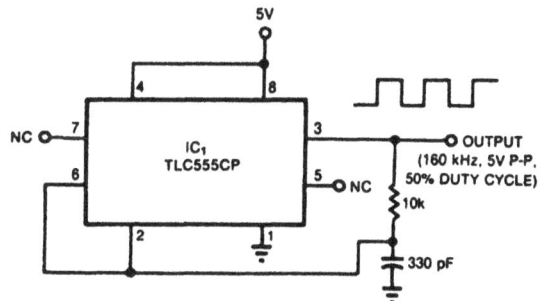

EDN

Fig. 5-25

HIGH-CURRENT OSCILLATOR

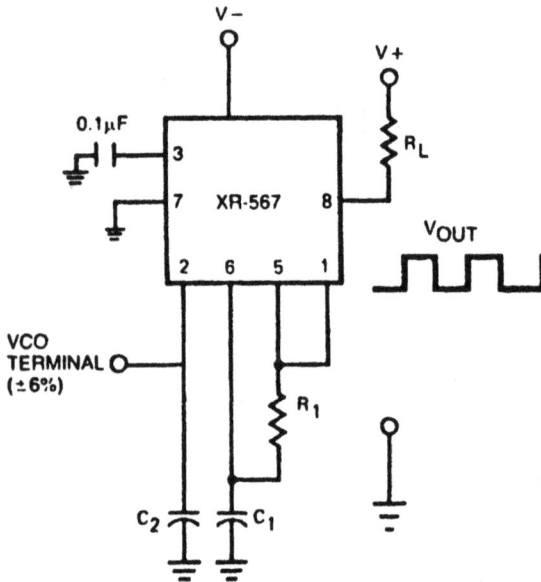

The oscillator output of the XR-567 can be amplified using the output amplifier and high-current logic output available at pin 8. In this manner, the circuit can switch 100-mA load currents without sacrificing oscillator stability. The oscillator frequency can be modulated over $\pm 6\%$ in frequency by applying a control voltage of pin 2.

EXAR

Fig. 5-26

QUADRATURE-OUTPUTS OSCILLATOR

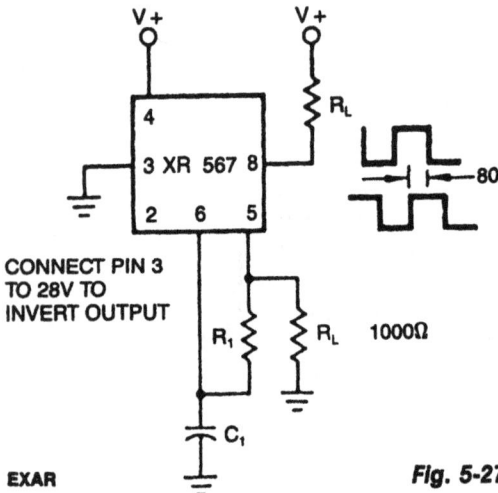

The XR-567 functions as a precision oscillator with two separate square-wave outputs at pins 5 and 8, that are at nearly quadrature phase with each other. Because of the internal biasing arrangement, the actual phase shift between the two outputs is typically 80%.

EXAR

Fig. 5-27

ASTABLE MULTIVIBRATOR

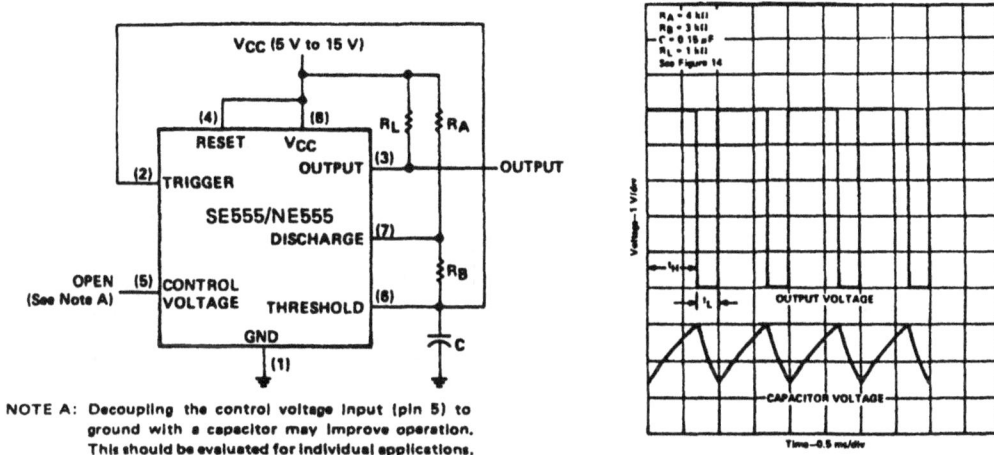

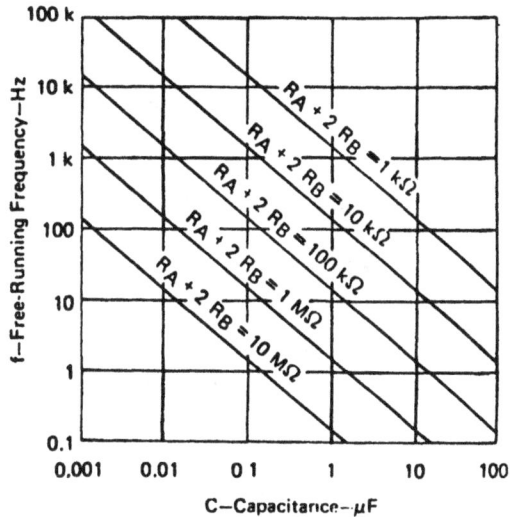

NOTE A: Decoupling the control voltage input (pin 5) to ground with a capacitor may improve operation. This should be evaluated for individual applications.

Fig. 5-28

The capacitor C will charge through R_A and R_B, and then discharge through R_B only. The duty cycle may be controlled by the values of R_A and R_B.

MONOSTABLE CIRCUIT

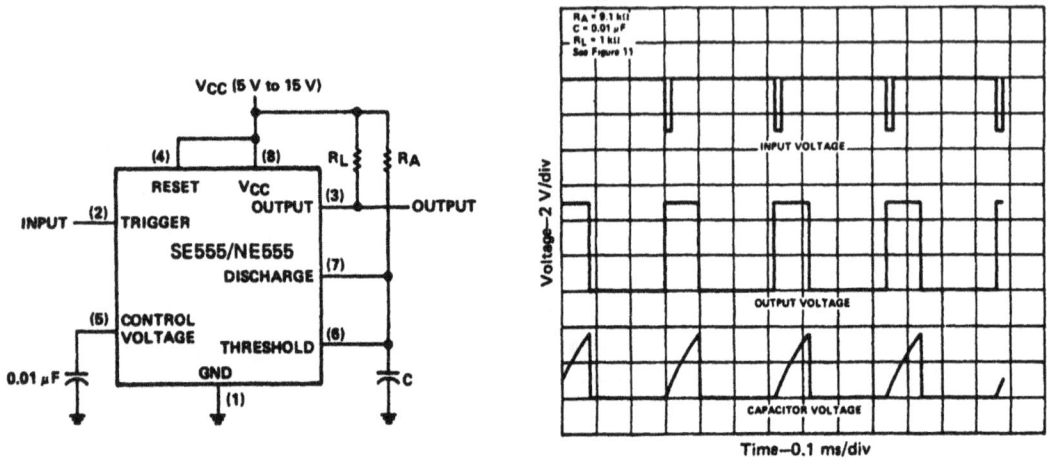

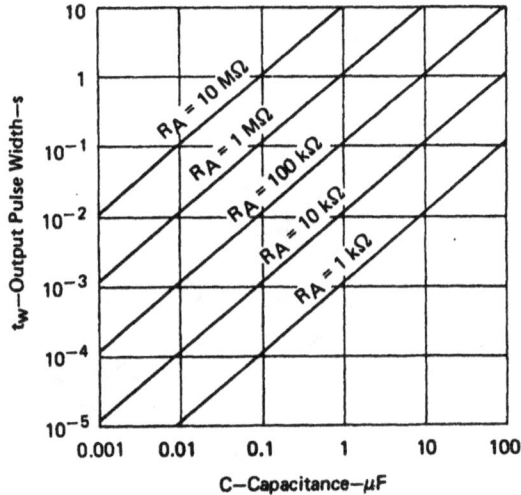

Time—0.1 ms/div

OUTPUT PULSE WIDTH vs CAPACITANCE

Fig. 5-29

If the output is low, application of a negative-going pulse to the trigger input sets the flip-flop (Q goes low), drives the output high, and turns off 1. Capacitor C is then charged through R_A until the voltage across the capacitor reaches the threshold voltage of the threshold input. If the trigger input has returned to a high level, the output of the threshold comparator will reset the flip-flop (Q goes high), drive the output low, and discharge C through Q1. Monostable operations is initiated when the trigger input voltage falls below the trigger threshold. Once initiated, the sequence will complete only if the trigger input is high at the end of the timing interval.

ASTABLE OSCILLATOR

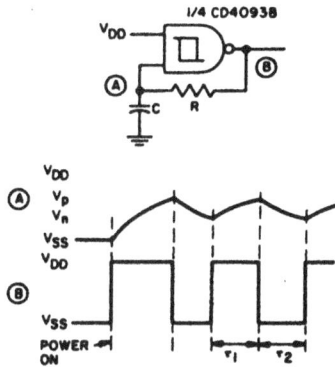

Before power is applied, the input and output are at ground potential and capacitor C is discharged. On power-on, the output goes high (V_{DD}) and C charges through R until V is reached; the output then goes low (V_{SS}). C is now discharged through R until V_n is reached. The output then goes high and charges C toward V_p through R. Thus, input A alternately swings between V_p and V_n as the output goes high and low. This circuit is self-starting at power-on.

RCA

Fig. 5-30

DIGITALLY CONTROLLED ASTABLE MULTIVIBRATOR

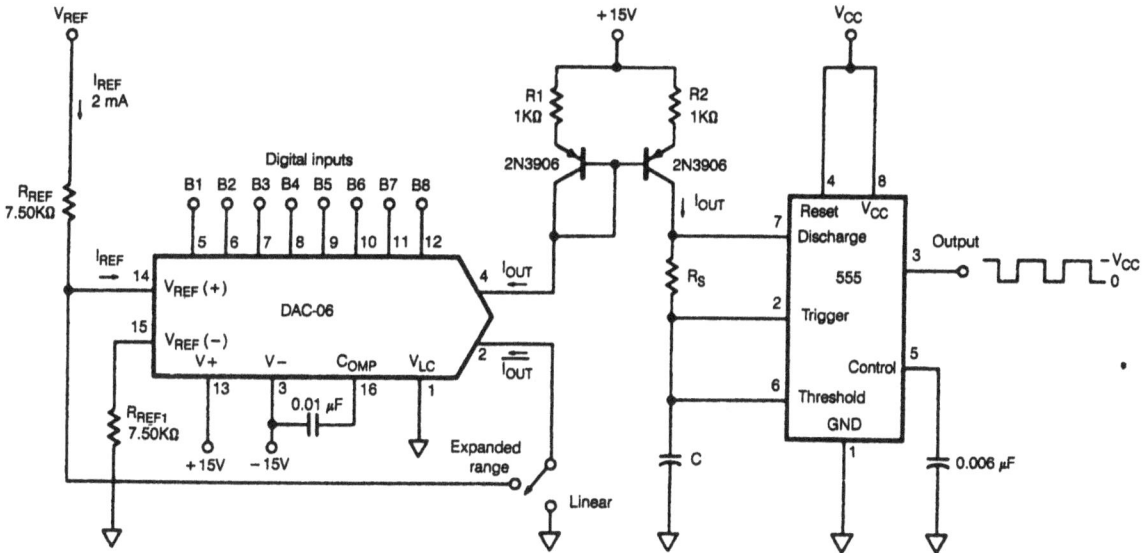

$$\text{FREQUENCY, } f = \cfrac{1}{\dfrac{1}{3}\dfrac{R_{REF}C}{(D)}\dfrac{V_{CC}}{V_{REF}} + 0.695\,R_BC}\quad \text{FOR LINEAR MODE}$$

$$\text{FREQUENCY, } f = \cfrac{1}{\dfrac{1}{3}R_{REF}C\dfrac{V_{CC}}{V_{REF}}\left[\dfrac{2-(D)}{(D)}\right] + 0.695\,R_BC}\quad \text{FOR EXPANDED MODE}$$

PRECISION MONOLITHICS

Fig. 5-31

6

RF Oscillators

The sources of the following circuits are contained in the Sources section, which begins on page 173. The figure number contained in the box of each circuit correlates to the source entry in the Sources section.

5-MHz VFO
Colpitts Oscillator
RF Oscillator
RF Genie
400-MHz Oscillator
2-MHz Oscillator

1.0-MHz Oscillator
Hartley Oscillator
500-MHz Oscillator
Low-Distortion Oscillator
Emitter-Coupled Big-Loop Oscillator
Wide-Range Oscillator

5-MHz VFO

VFO

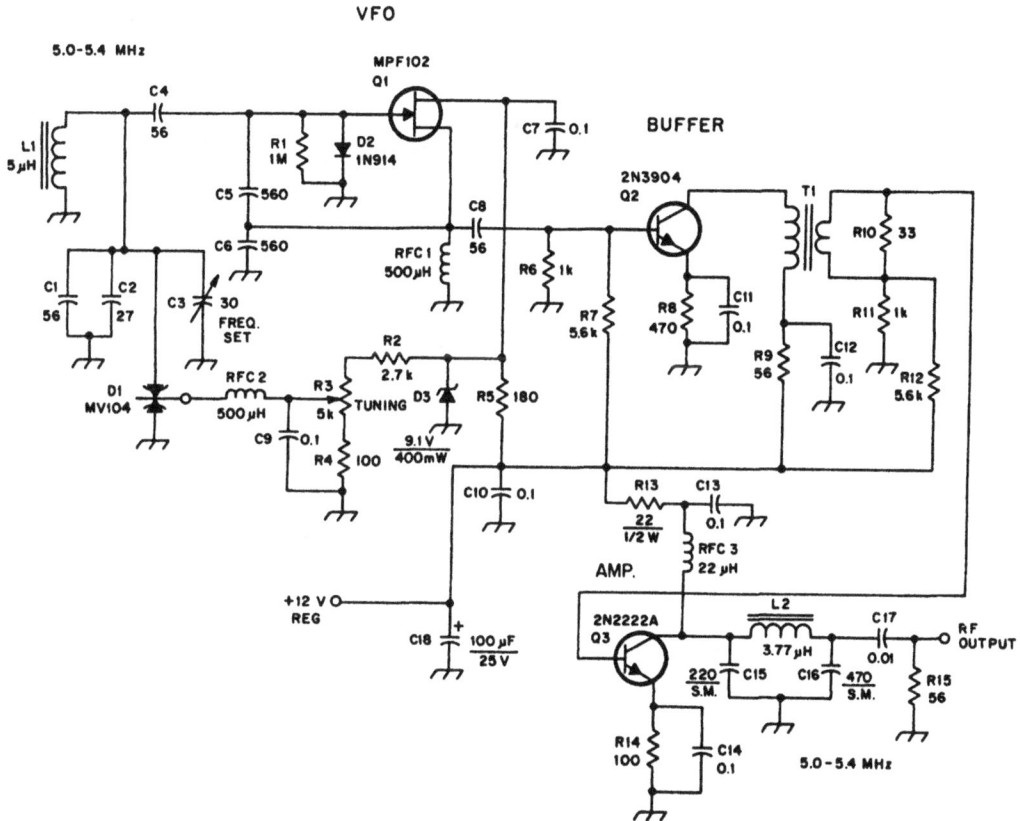

5.0-5.4 MHz

L1 5 μH

C4 56

MPF102 Q1

R1 1M

D2 1N914

C7 0.1

BUFFER

2N3904 Q2

T1

R10 33

C5 560

C8 56

RFC1 500μH

C6 560

R6 1k

R8 470

C11 0.1

R11 1k

C1 56

C2 27

C3 30 FREQ. SET

R7 5.6k

R9 56

C12 0.1

R12 5.6k

D1 MV104

RFC2 500 μH

R3 5k TUNING

R2 2.7k

D3

R5 180

C9 0.1

R4 100

9.1V 400mW

C10 0.1

R13 22 1/2 W

C13 0.1

RFC 3 22 μH

AMP.

+12 V REG

C18 100 μF 25V

2N2222A Q3

L2 3.77μH

C17 0.01

RF OUTPUT

220 S.M. C15

C16 470 S.M.

R15 56

R14 100

C14 0.1

5.0-5.4 MHz

Fig. 6-1

A JFET (Q1) serves as the oscillator. D2 helps to stabilize the transistor by limiting positive sine wave peaks and stabilizing the bias. Output from Q1 is supplied to a Class A buffer, Q2. It operates as a broadband amplifier by means of T1, which is untuned. Output amplifier Q3 is also a Class A stage. A low-pass, single-section filter is used at the output of Q3 to remove some of the harmonic currents generated within the system. The filter output impedance is 50 Ω. The injection level to the mixer is 600 mV p-p.

COLPITTS OSCILLATOR

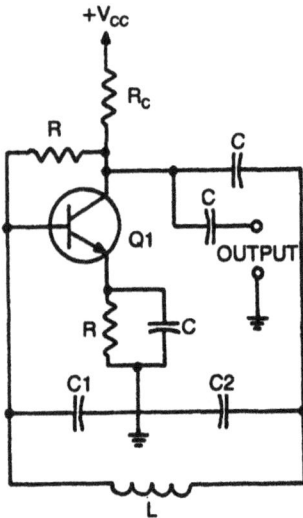

When calculating its resonant frequency, use C1C2/C1 + C2 for the total capacitance of the L-C circuit.

Fig. 6-2

RF OSCILLATOR

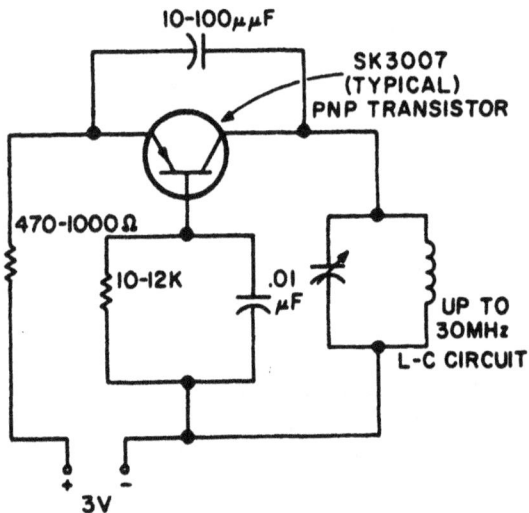

This RF oscillator is useful up to 30 MHz. An SK 3007 pnp transistor is recommended.

Fig. 6-3

RF GENIE

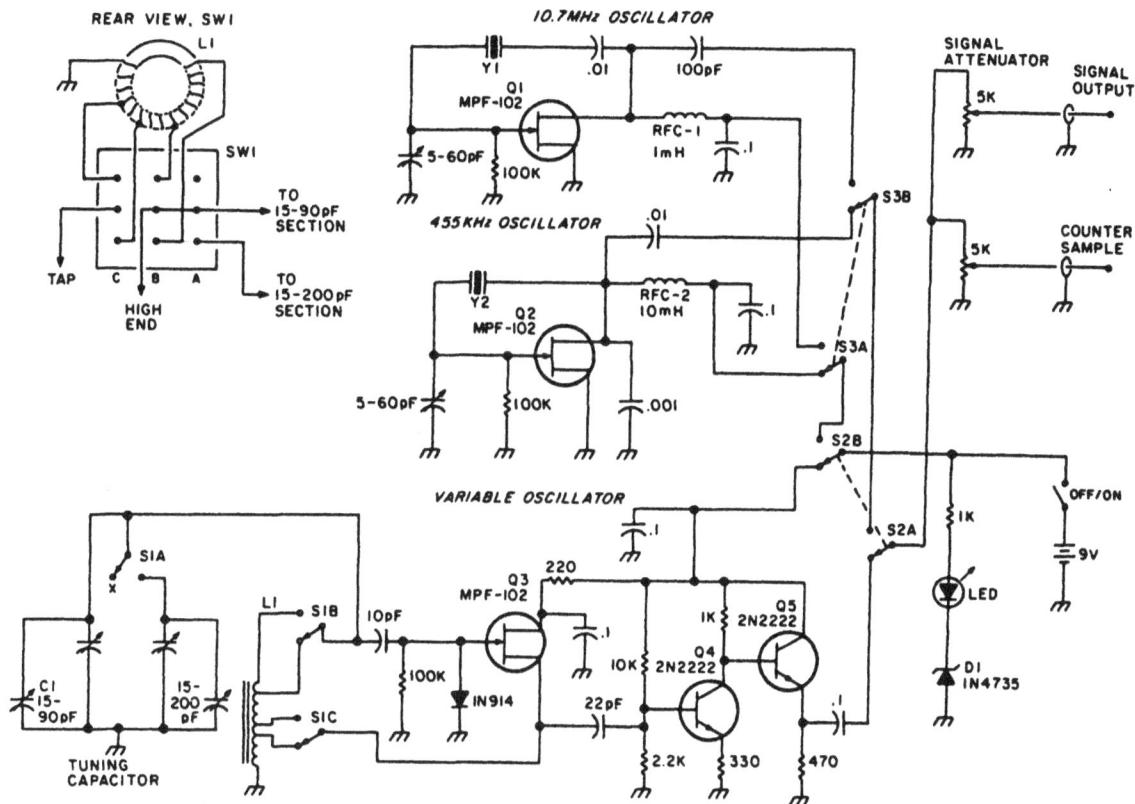

Fig. 6-4

A variable oscillator covers 3.2 to 22 MHz in two bands—providing coverage of 80 through 15 meters plus most crystal-filter frequencies. Optional 455-kHz and 10.7-MHz crystal oscillators can be switched on-line for precise IF alignment. Generator output is on the order of 4 volts p-p into a 500-Ω load. A simple voltage-divider attenuator controls the generator's output level, and a second output provides sufficient drive for an external frequency counter.

400-MHz OSCILLATOR

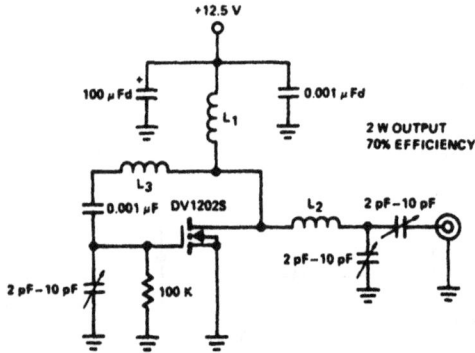

+12.5 V

100 µFd   L₁   0.001 µFd

2 W OUTPUT
70% EFFICIENCY

L₃
0.001 µF   DV1202S   L₂   2 pF–10 pF

2 pF–10 pF

2 pF–10 pF   100 K

Parts List

L₁—8 turns #22 closewound on 1/4" diameter
L₂—1/2 inch #16 wire
L₃—1 inch #16 wire

SILICONIX   *Fig. 6-5*

1.0-MHz OSCILLATOR

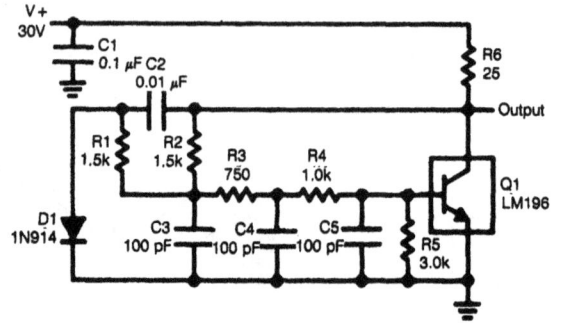

V+
30V   C1
0.1 µF C2
0.01 µF

R6
25

Output

R1
1.5k   R2
1.5k   R3
750   R4
1.0k   Q1
LM196

D1
1N914   C3
100 pF   C4
100 pF   C5
100 pF   R5
3.0k

NATIONAL SEMICONDUCTOR   *Fig. 6-7*

2-MHz OSCILLATOR

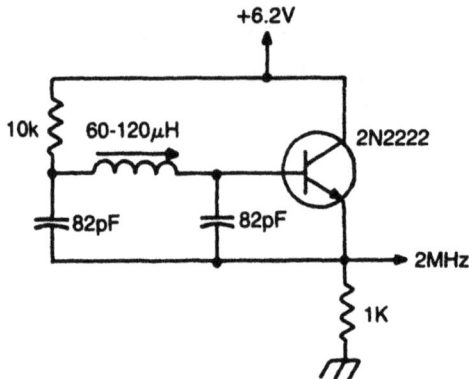

+6.2V

10k   60-120µH   2N2222

82pF   82pF

2MHz

1K

TAB BOOKS   *Fig. 6-6*

Miller 9055 miniature slugtuned coil; all resistors 1/4W 5%; all caps min. 25 V ceramic.

HARTLEY OSCILLATOR

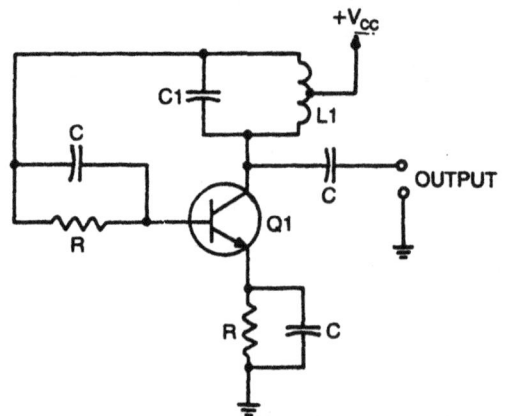

+V$_{cc}$

C1   L1

C   C

R   Q1   OUTPUT

R   C

RADIO-ELECTRONICS   *Fig. 6-8*

Resonant frequency is $1/2 \, \pi \, \sqrt{L1C1}$.

500-MHz OSCILLATOR

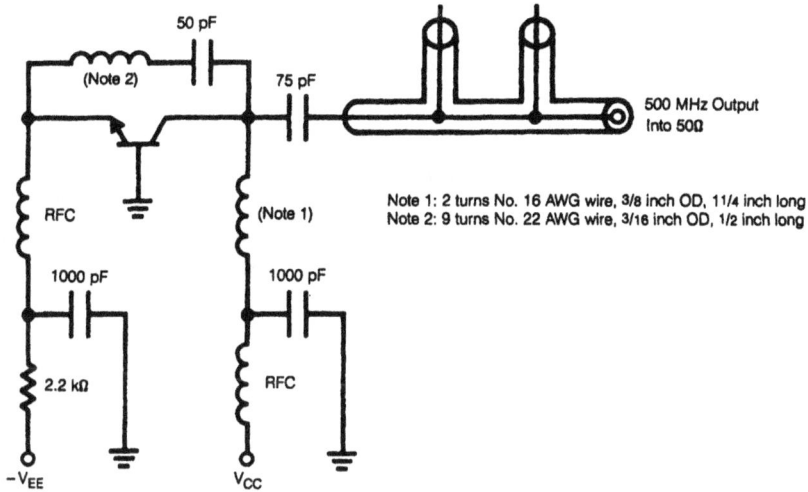

Note 1: 2 turns No. 16 AWG wire, 3/8 inch OD, 1¼ inch long
Note 2: 9 turns No. 22 AWG wire, 3/16 inch OD, 1/2 inch long

NATIONAL SEMICONDUCTOR

Fig. 6-9

LOW-DISTORTION OSCILLATOR

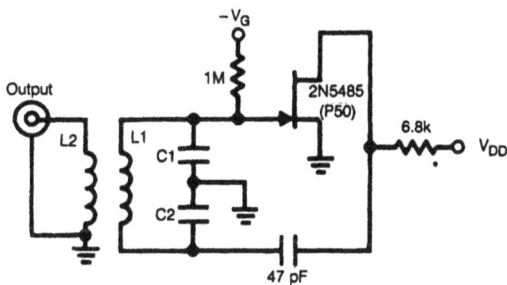

20 MHz oscillator values

C1 ≈ 700 pF	L1 = 1.3 μH
C2 = 75 pF	L2 = 10T 3/8" dia 3/4" long
V_{DD} = 16V	I_D = 1 mA

20 MHz oscillator performance

Low distortion 20 MHz osc
2nd harmonic − 60 dB
3rd harmonic > −70 dB

NATIONAL SEMICONDUCTOR

Fig. 6-10

The 2N5485 JFET is capable of oscillating in a circuit where harmonic distortion is very low. The JFET local oscillator is excellent when a low harmonic content is required for a good mixer circuit.

EMITTER-COUPLED BIG-LOOP OSCILLATOR

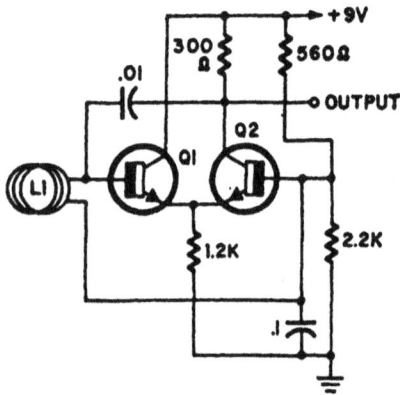

L1 is a loop of 10 to 20 turns of insulated wire with a diameter anywhere between 4″ to 4′. This oscillator frequency (7 to 30 MHz) shifts substantially when a person comes near or into the loop. This oscillator together with a resonant detector might make a very good antipersonnel alarm. Transistors are 2N2926 or equivalent.

RADIO-ELECTRONICS

Fig. 6-11

WIDE-RANGE OSCILLATOR

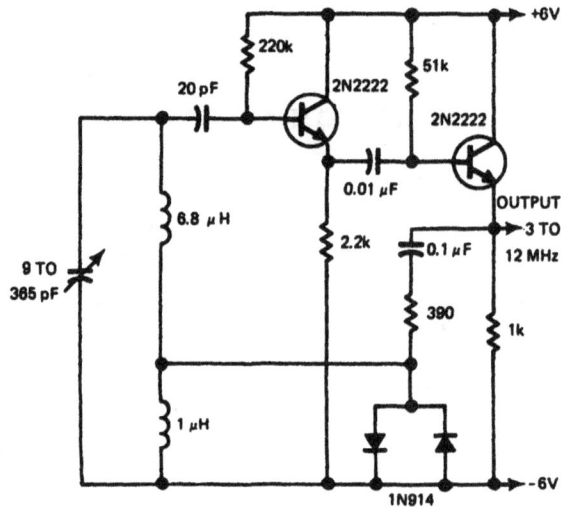

EDN

Fig. 6-12

The gain control allows the oscillator to maintain essentially constant output over its range. The circuit functions over 160 kHz to 12 MHz with essentially constant amplitude.

7

Sirens, Warblers and Wailers

The sources of the following circuits are contained in the Sources section, which begins on page 173. The figure number contained in the box of each circuit correlates to the source entry in the Sources section.

TWO-TONE SIREN

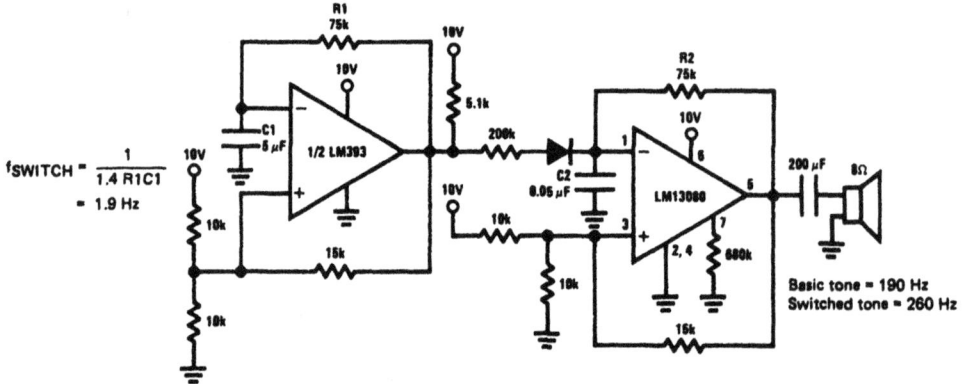

$$f_{SWITCH} = \frac{1}{1.4 \, R1C1}$$
$$= 1.9 \text{ Hz}$$

Basic tone = 190 Hz
Switched tone = 260 Hz

NATIONAL SEMICONDUCTOR

Fig. 7-1

This siren provides a constant audio output, but alternates between two separate tones. The LM13080 is set to oscillate at one basic frequency; this frequency is changed by adding a 200-kΩ charging resistor in parallel with the feedback resistor, R2.

YELPING SIREN

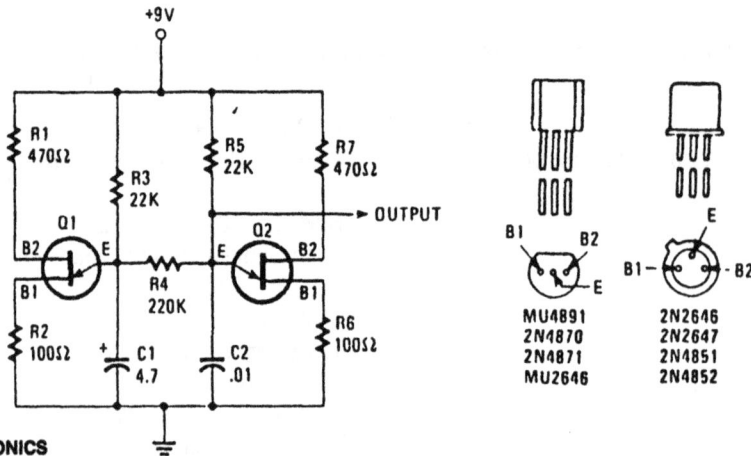

MU4891
2N4870
2N4871
MU2646

2N2646
2N2647
2N4851
2N4852

HANDS-ON ELECTRONICS

Fig. 7-2

Unijunction transistors Q1 and Q2 are both connected as relaxation-type, sawtooth oscillators. Transistor Q1 is the low-frequency control oscillator and Q2 is the tone generator. Sawtooth waveforms are produced at the emitter terminals. Without R4 connecting the two emitters, each oscillator operates independently, with its frequency determined mainly by the RC time constant. With the values shown, Q1 operates from 1 to 1.5 Hz and Q2 operates from 400 to 500 Hz. When R4 is connected between the two emitters, it couples the low-frequency sawtooth from Q1 directly across capacitor C2. That coupling causes the frequency of the tone generator to increase, along with the rise in sawtooth voltage from Q1. The tone generator's frequency drops to its lower design value when C1 discharges and produces the falling edge of the sawtooth.

SIREN

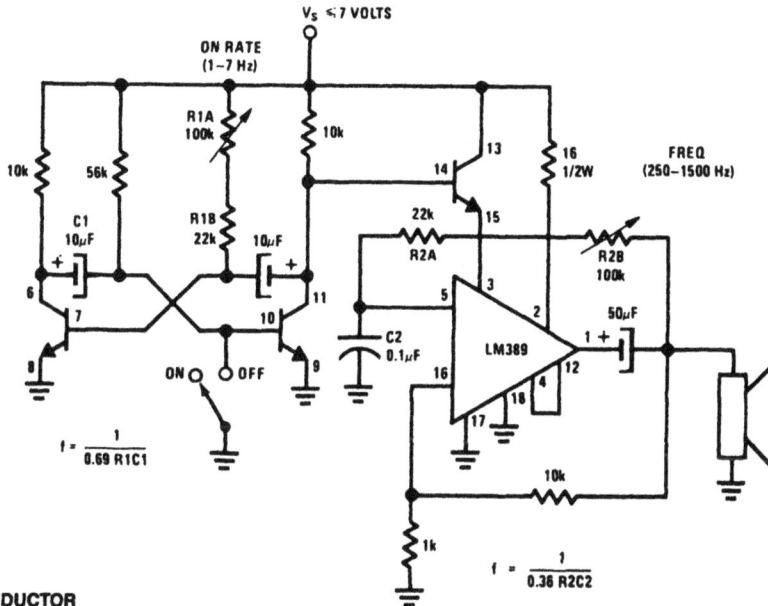

$$f = \frac{1}{0.69\ R1C1}$$

$$f = \frac{1}{0.36\ R2C2}$$

Fig. 7-3

This circuit uses one of the LM389 transistors to gate the power amplifier on and off by applying the muting technique. The other transistors form a cross-coupled multivibrator circuit that controls the rate of the square-wave oscillator. The power amplifier is used as the square-wave oscillator with individual frequency adjust provided by potentiometer R2B.

TWO-STATE SIREN

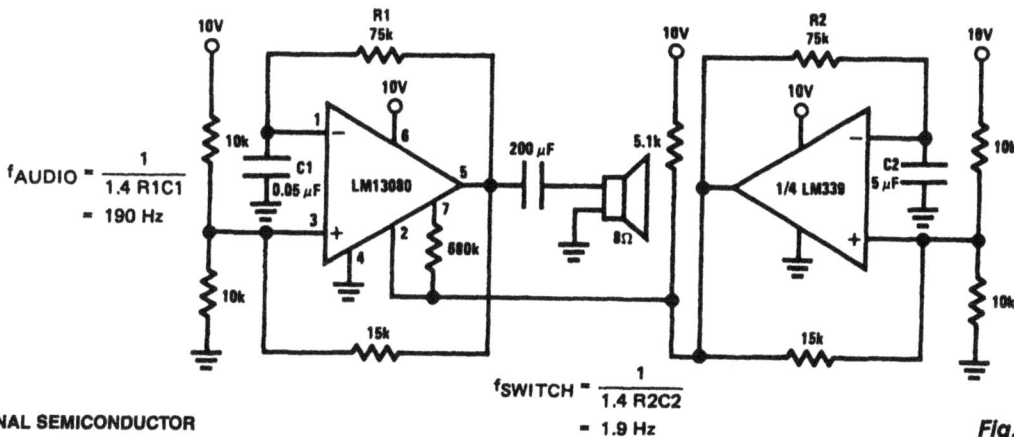

$$f_{AUDIO} = \frac{1}{1.4\ R1C1}$$
$$= 190\ Hz$$

$$f_{SWITCH} = \frac{1}{1.4\ R2C2}$$
$$= 1.9\ Hz$$

NATIONAL SEMICONDUCTOR

Fig. 7-4

This is a two-state or on/off-type siren where the LM13080 oscillates at an audio frequency and drives an 8-Ω speaker. The LM339 acts as a switch, which controls the audio burst rate.

7400 SIREN

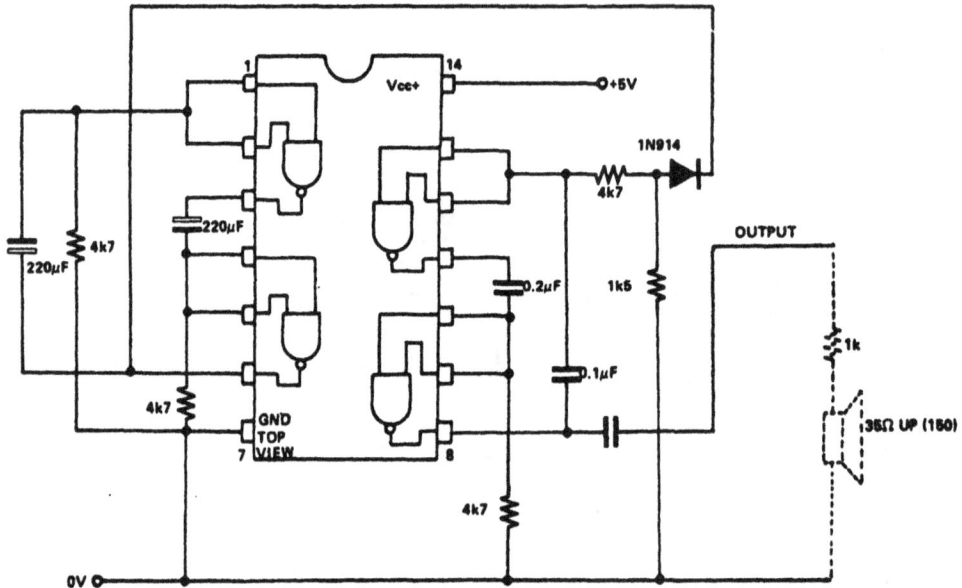

Fig. 7-5

Two NAND gates are used for the oscillator, and two as the control. If the two-tone speed needs to be altered, the 220-μF capacitors can be changed (larger for slower operation). If the frequency of the oscillator is to be changed, the 0.2- and 0.1-μF capacitors can be varied and the value of R1 can be increased. To change frequency range between the two notes, alter the 1.5-kΩ resistor.

TOY SIREN

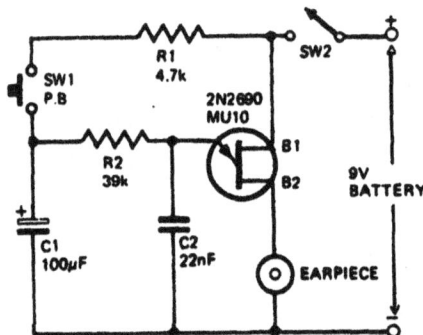

Fig. 7-6

This circuit can be built small enough to fit inside a toy. The circuit consists of a relaxation oscillator utilizing one unijunction transistor (2N2646, MU10, TIS43). R2 and C2 determine the frequency of the tone. Pushing the button, SW1 charges up the capacitor and the potential at the junction of R2 and C2 rises, causing an upswing in the oscillation frequency. On releasing the pushbutton, the charge on C2 will drop slowly with a proportional reduction in the frequency of oscillation. Manual operation of the button at intervals of approximately 2 seconds will produce a siren sound.

MULTIFUNCTION SIREN SYSTEM

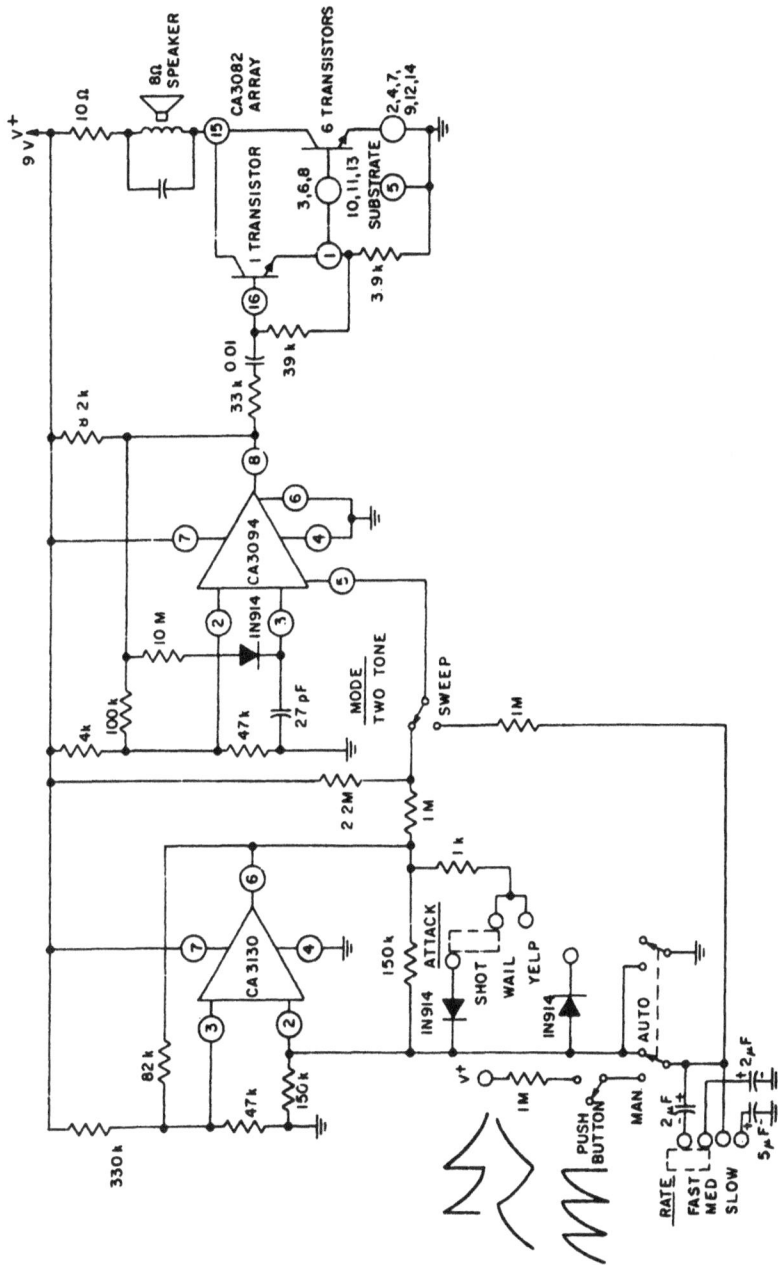

GENERAL ELECTRIC/RCA

Fig. 7-7

The circuit uses a CA3130 BiMOS op amp as a multivibrator to control the siren's rate. A CA3094 used as a VCO is followed by a CA3082 transistor array to drive a speaker. A "Manual" or "Auto" mode switch allows the user to select either intermittent or continuous siren operation, respectively. In addition, three switches are available to control "Mode," "Attack," and "Rate."

153

LINEAR IC SIREN

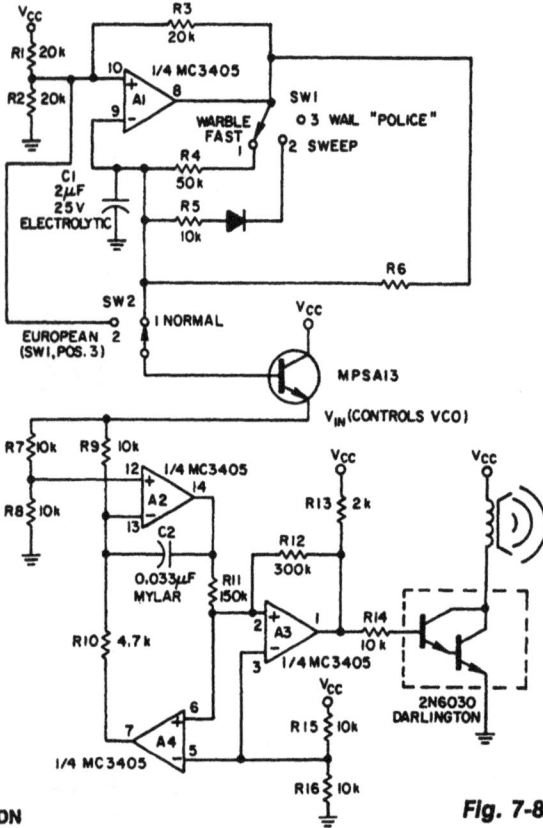

A low-frequency, op-amp oscillator and a VCO, both configured from a single MC3405 dual op amp and dual comparator, are the major components in a siren circuit that can be made to produce various warbles and wails, or serve as an audio sweep generator. The only other active components needed are an MPS A13 small-signal transistor and a 2N6030 power Darlington transistor.

EDN

Fig. 7-8

SUPER SOUND GENERATOR

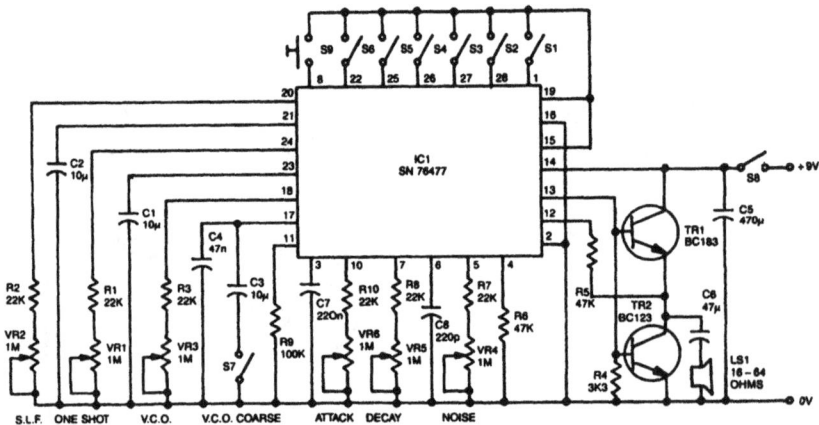

EVERYDAY ELECTRONICS

Fig. 7-9

SUPER SOUND GENERATOR (continued)

Six preset controls and seven selector switches enable a vast range of different sounds to be produced and altered at will. Such sounds as steam trains chuffing, helicopters flying, birds chirping, and machine guns firing are possible, as well as the usual police sirens, phaser guns, and bomb explosions. The circuit incorporates an amplifier giving 150-mW output into a small loudspeaker. Alternatively, a separate amplifier system can be used for disco effects, car alarms, etc. Continuous or one-shot sounds are possible. For one-shot sounds, a pushbutton switch is provided, which can also be used to turn continuous sounds on and off. A single IC, SN76477, provides all of the sound generation circuits.

HEE-HAW SIREN

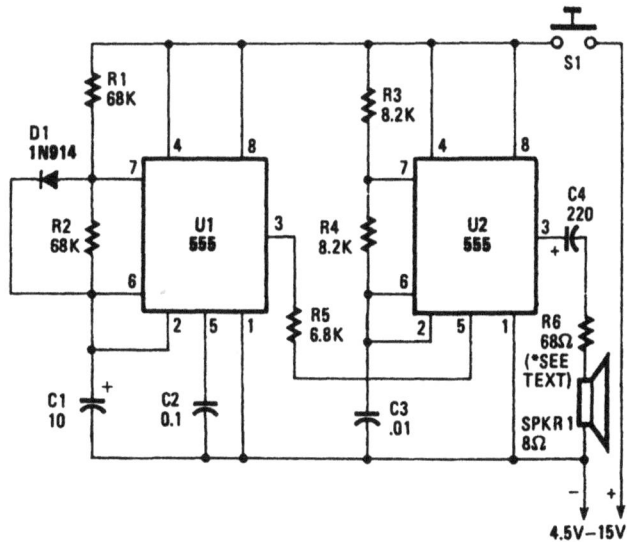

Fig. 7-10

A pair of timer IC's are the heart of a circuit that simulates the warbling hee-haw of a British police siren. One of the 555 timers, U2, is wired as an astable multivibrator operating at about 900 Hz. The other, U1, operates at approximately 1 Hz. Its output at pin 3 is a square wave with a 50% duty cycle—on and off cycles of about 0.5 second each. The output of U1 is applied to pin 5, the control-voltage terminal of U2. The frequency of the 555 timer IC is relatively independent of supply voltage, but can be varied over a fairly wide range by applying a variable voltage between pin 5 and ground. When U1's output becomes high, U2 operates at about 800 kHz. That switching between two frequencies produces the warbling hee-haw signal.

PROGRAMMABLE-FREQUENCY ADJUSTABLE-RATE SIREN

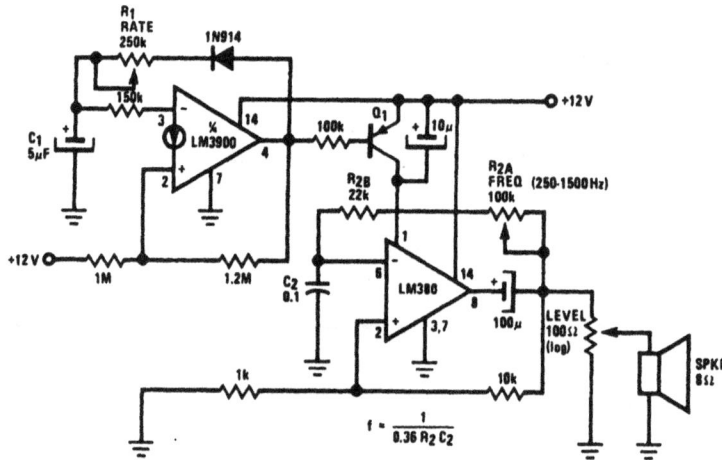

$$f = \frac{1}{0.36\,R_2\,C_2}$$

NATIONAL SEMICONDUCTOR

Fig. 7-11

The LM380 operates as an astable oscillator with the frequency determined by R2/C2. Adding Q1 and driving its base, with the output of an LM3900 wired as a second astable oscillator, acts to gate the output of the LM380 on and off, at a rate fixed by R1/C1.

THE WAILING SIREN

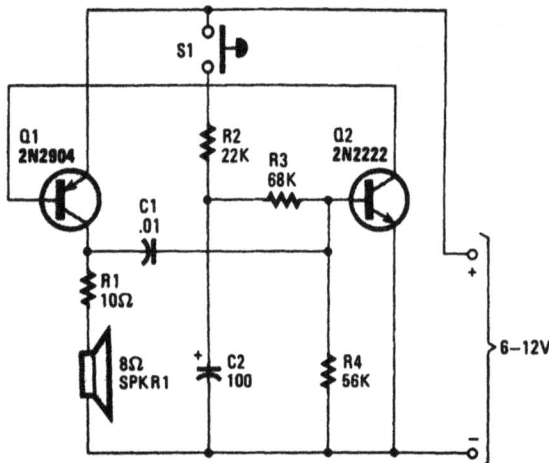

HANDS-ON ELECTRONICS

Fig. 7-12

Transistors Q1 and Q2, with feedback provided via C1 from the collector of Q1 to the base of Q2, forms a voltage-controlled oscillator (VCO). De-pending on the voltage applied to Q2's base, the VCO frequency ranges from around 60 Hz to 7.5 kHz. The instantaneous voltage applied to the base of Q2 is determined by the values of C2, R2, R3, and R4. When pushbutton switch S1 is closed, C2 charges fairly rapidly to the maximum supply voltage through R2, a 22-kΩ fixed resistor. That causes the siren sound to rise rapidly to its highest frequency. When the button is released, the capacitor discharges through R3 and R4 with a combined resistance of 124 kΩ, causing the siren sound to decay from a high-pitched wail to a low growl. If you want to experiment with the pitch of the sound at its highest frequency, try different values for C1. Increase its value for lower notes, and decrease it for higher ones. Different values for R2 will change the attack time. A 100-kΩ resistor provides equal attack and decay times. The way you handle the pushbutton varies the effect.

HIGH-POWER SIREN

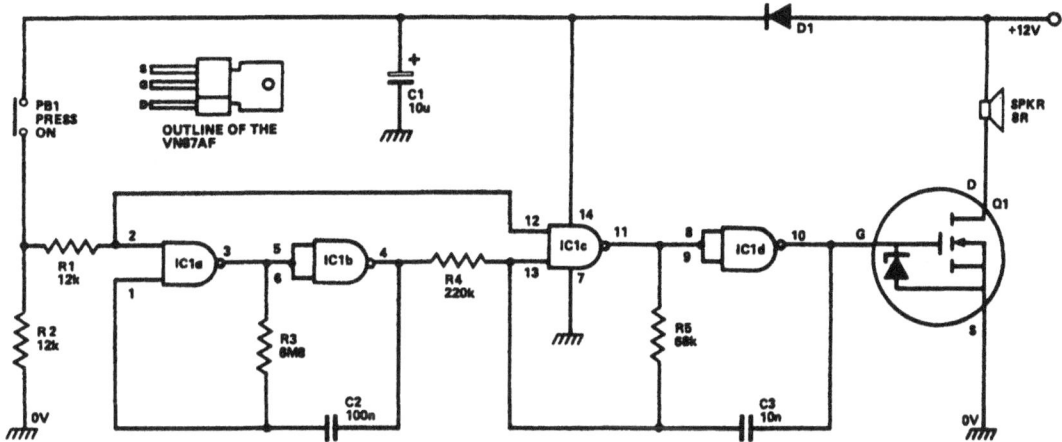

ELECTRONICS TODAY INTERNATIONAL

NOTES:
D1 IS 1N4001
Q1 IS VN67AF
IC1 IS CD4011B

Fig. 7-13

IC1a and IC1b are wired as a slow astable multivibrator and IC1c-IC1d are wired as a fast astable. Both are "gated" types, which can be turned on and off via PB1. The output of the slow astable modulates the frequency of the fast astable, and the output of the fast astable is fed to the external speaker via the Q1 VMOS power-FET amplifier stage.

HEE-HAW TWO-TONE SIREN

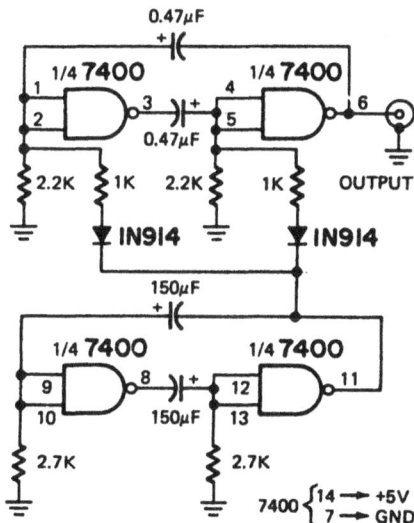

RADIO-ELECTRONICS *Fig. 7-14*

The circuit uses two gates of a 7400 IC, which is cross-connected to form an astable multivibrator driven by the 1-pulse per second output of the digital clock IC. The hee-haw circuit has a low-frequency astable modulator added to make a self-contained European-type siren. Tone and rate can be varied as desired by changing capacitor values. If the tone is too harsh, a simple R-C filter will remove the harmonic content—the multivibrator output is almost a square wave. With the resistor values shown, no start-up problems occur; but if the 2.2 or 2.7 kΩ resistors are changed too much, latch-up can be a problem.

SIREN SIMULATES STAR TREK RED ALERT

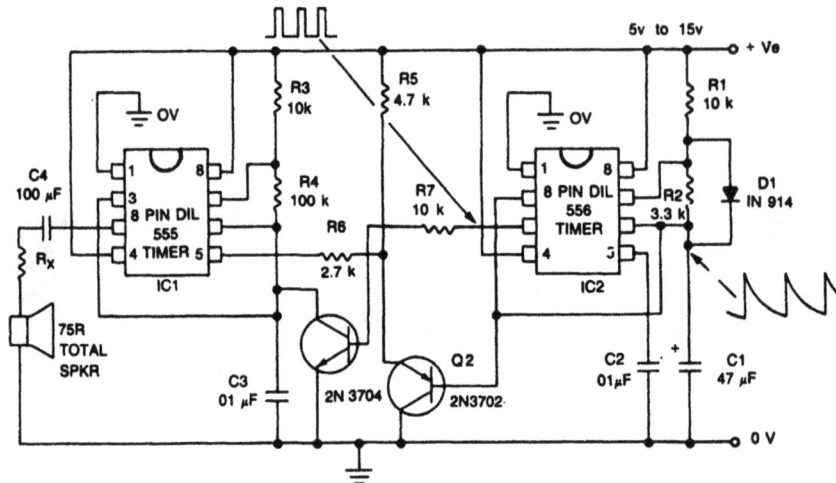

ELECTRONICS TODAY INTERNATIONAL

Fig. 7-15

The signal starts at a low frequency, rises for about 1.15 seconds to a high frequency, ceases for about 0.35 seconds, then starts rising again from a low frequency, and so on, ad infinitum.

YELP OSCILLATOR

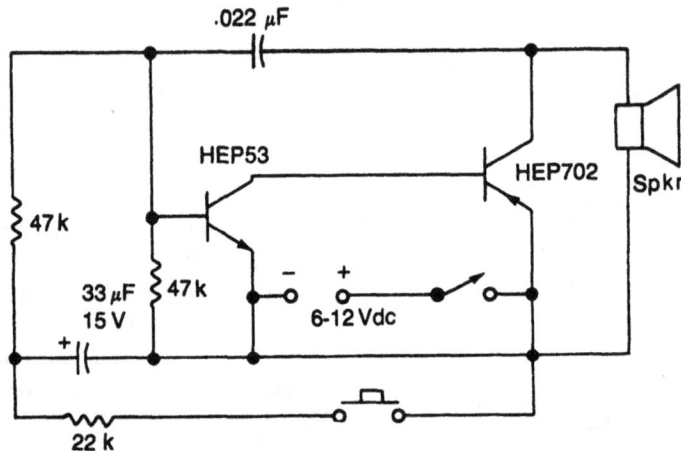

WILLIAM SHEETS

Fig. 7-16

Close the pushbutton switch and the circuit starts the siren up-shifting to a higher frequency. Release it and the tone slides down until S2 is closed again. Tone quality is adjusted by changing the 0.022-μF capacitor.

ELECTRONIC SIREN

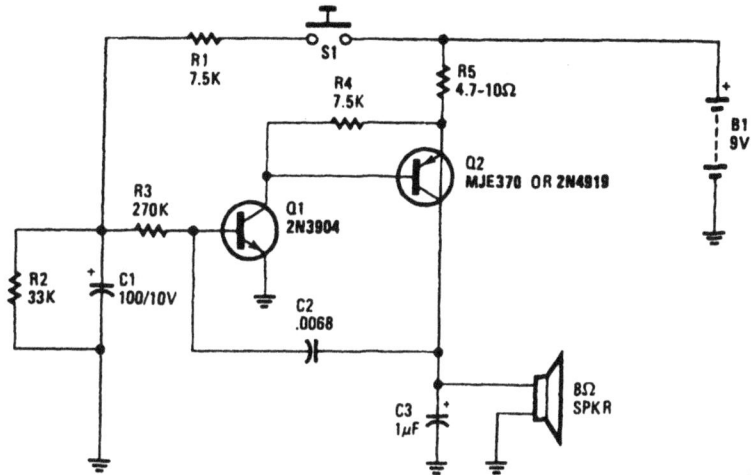

Fig. 7-17

Reprinted with permission from Radio-Electronics Magazine, December 1981. Copyright Gernsback Publications, Inc., 1981.

The wailing sound of a siren is generated by a VFO consisting of Q1 and Q2. Capacitor C2 provides the feedback for the oscillator. The frequency of the oscillator is varied by the voltage applied to the base of Q1 through R3. When switch S1 is closed, capacitor C1 charges, thus increasing the oscillator frequency. When S1 is released, capacitor C1 discharges, and the oscillator frequency decreases. Capacitor C3 limits the maximum oscillator frequency. The average battery current drain is about 15 mA.

555 BEEP TRANSFORMER

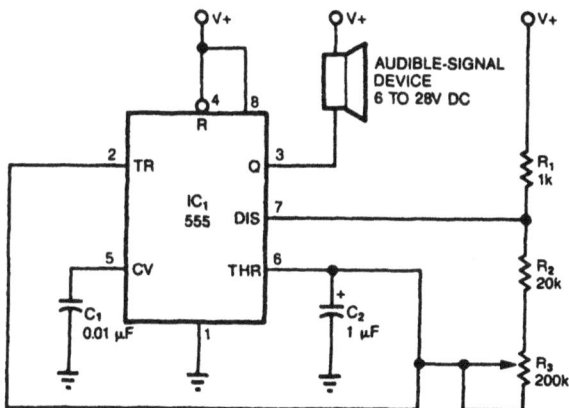

The simple circuit transforms the steady beep of an audible-signal device, such as a Mallory sonalert, into a distinctive warble or chirp. The value of C2 determines just what tone color you'll get. With the 1-μF value shown, the circuit produces a warble similar to the ring tone of an inexpensive phone. A 10-μF value produces a chirp similar to a truck's back-up alarm. One elaboration of this circuit would be to use the second section of a 555 timer to drive a piezoelectric transducer instead of a sonalert; that modification would vary the tone's pitch, as well as the chirp rate.

ELECTRONIC DESIGN

Fig. 7-18

159

WARBLE-TONE ALARM

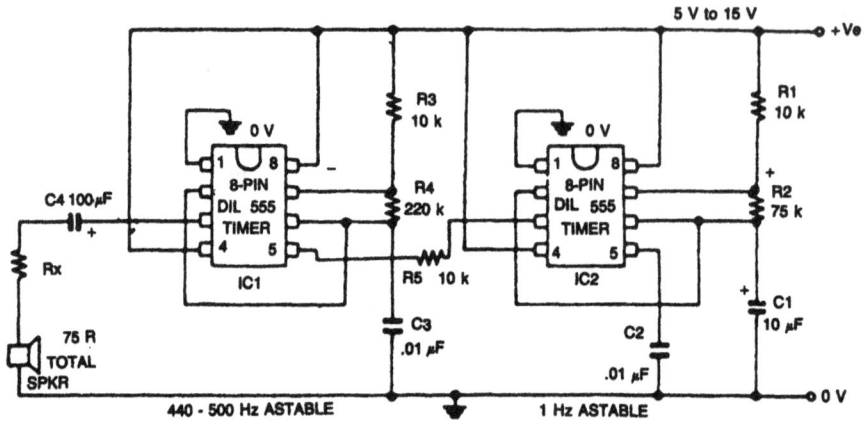

ELECTRONICS TODAY INTERNATIONAL

Fig. 7-19

The circuit generates a warble-tone alarm signal that simulates the sound of a British police siren. IC1 is wired as an alarm generator and IC2 is wired as a 1-Hz astable multivibrator. The output of IC2 is used to frequency modulate IC1 via R5. The action is such that the output frequency of IC1 alternates symmetrically between 500 Hz and 440 Hz, taking one sound to complete each alternating cycle.

WARBLING TONE GENERATOR

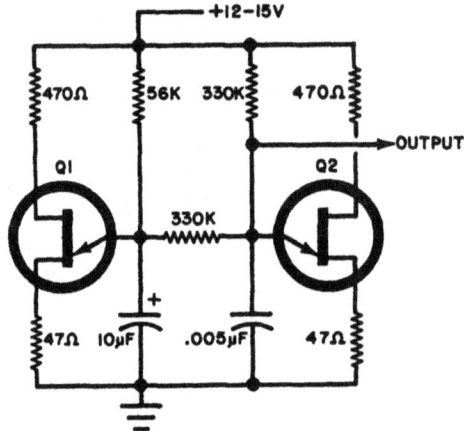

POPULAR ELECTRONICS

Fig. 7-20

The circuit uses two unijunction transistors. The low-frequency sawtooth generated by Q1 modulates the high-frequency tone generated by Q2. The output should feed into a high-impedance amplifier. Q1 = Q2 = 2N4871.

WARBLE GENERATOR

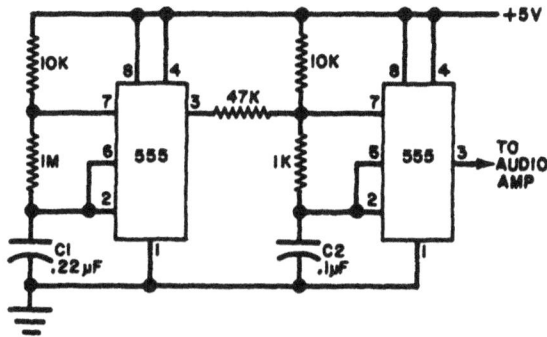

POPULAR ELECTRONICS

Fig. 7-21

The circuit uses a pair of 555 timers or a single dual timer. Capacitor C1 controls the speed of the warble, while C2 determines the pitch. The values shown should produce quite a distinctive signal.

WAILING ALARM

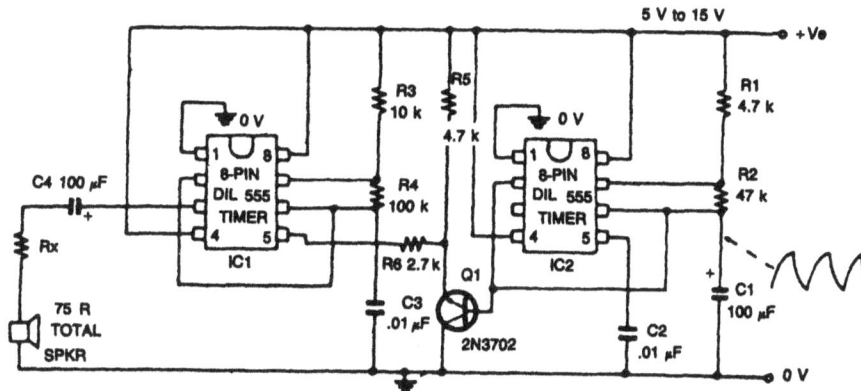

ELECTRONICS TODAY INTERNATIONAL

Fig. 7-22

This circuit simulates the sound of an American police siren. IC2 is wired as a low frequency astable that has a cycling period of about 6 seconds. The slowly varying ramp waveform on C1 is fed to pnp emitter-follower Q1, and is then used to frequency modulate alarm generator IC1 via R6. IC1 has a natural center frequency of about 800 Hz. Circuit action is such that the alarm output signal starts at a low frequency, rises for 3 seconds to a high frequency, then falls over 3 seconds to a low frequency again, and so on, ad infinitum.

SIREN USES TTL GATES

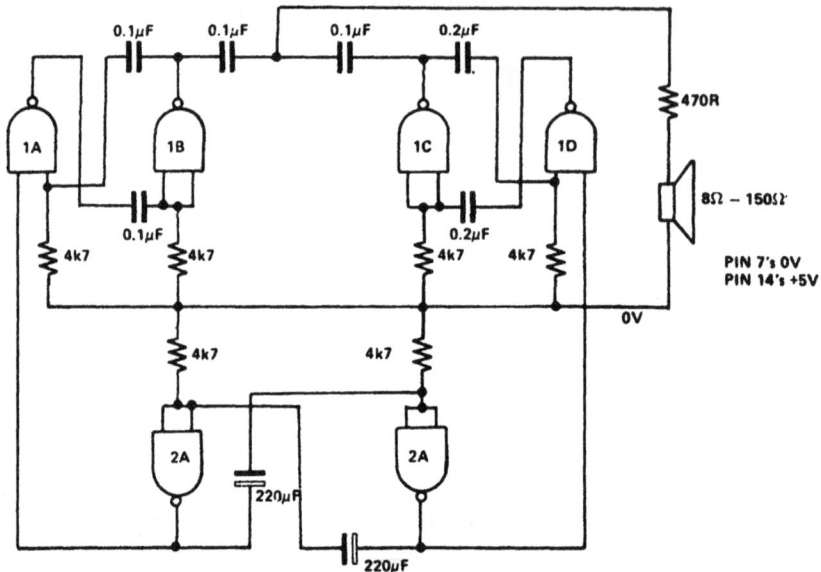

ELECTRONICS TODAY INTERNATIONAL

Fig. 7-23

The siren consists of two oscillator that generate the tones. A third oscillator is used to switch the others on and off alternately, giving the two-tone effect. By changing the capacitor values different tones can be produced.

ELECTRONIC SHIP SIREN

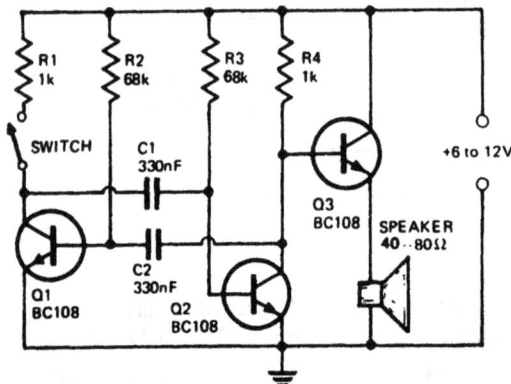

The circuit consists of a multivibrator (Q1 & Q2), and a low-power output stage Q3. The speaker should have an impedance in the region of 40 to 80 ohms. To use a low-impedance speaker, connect an output transformer from the emitter of Q3 to ground. C1 and C2 determine the pitch of the siren and the values specified will provide a tone of about 300 Hz. Quiescent current is negligible. The output at the collector of Q2 can also be fed into an amplifier input via a 1-μF electrolytic, in series with a 12-kΩ resistor.

ELECTRONICS TODAY INTERNATIONAL   *Fig. 7-24*

162

VARYING-FREQUENCY WARNING ALARM

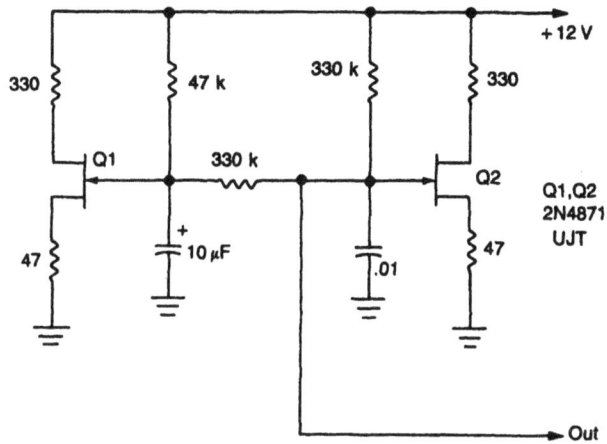

Fig. 7-25

The output frequency changes continuously. Low-frequency oscillator (Q1) modulates high frequency oscillator Q2 and its associated timing capacitor.

STEAM TRAIN WITH WHISTLE

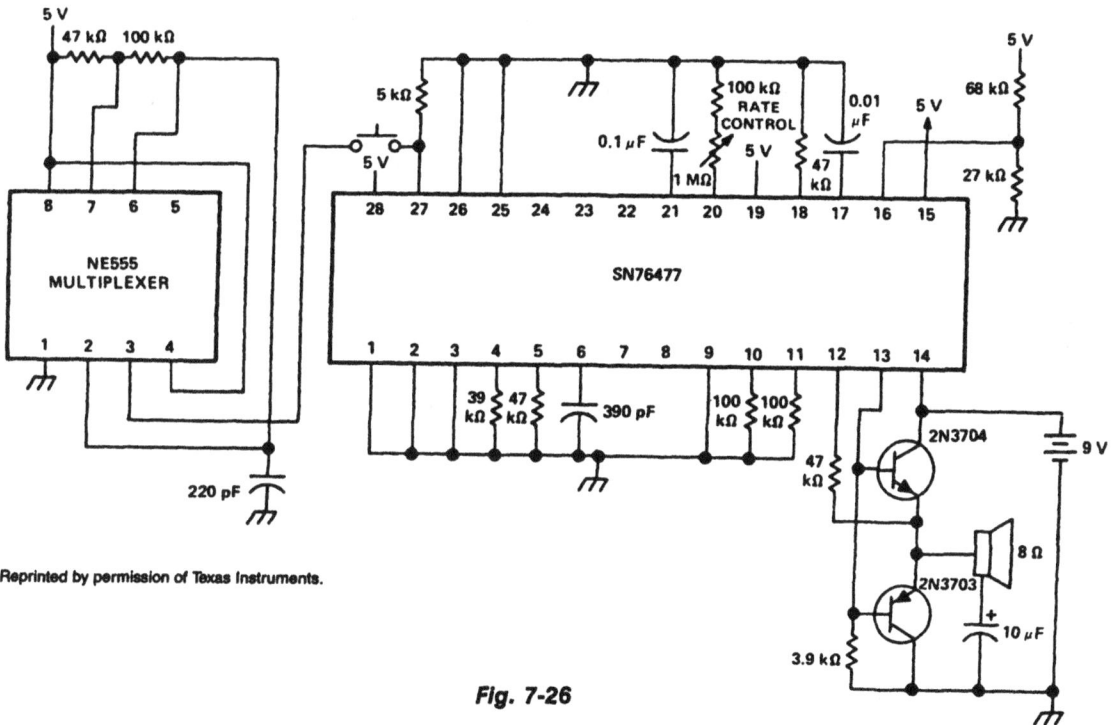

Fig. 7-26

8

Voltage-Controlled Oscillators

The sources of the following circuits are contained in the Sources section, which begins on page 173. The figure number contained in the box of each circuit correlates to the source entry in the Sources section.

LINEAR VOLTAGE-CONTROLLED OSCILLATOR

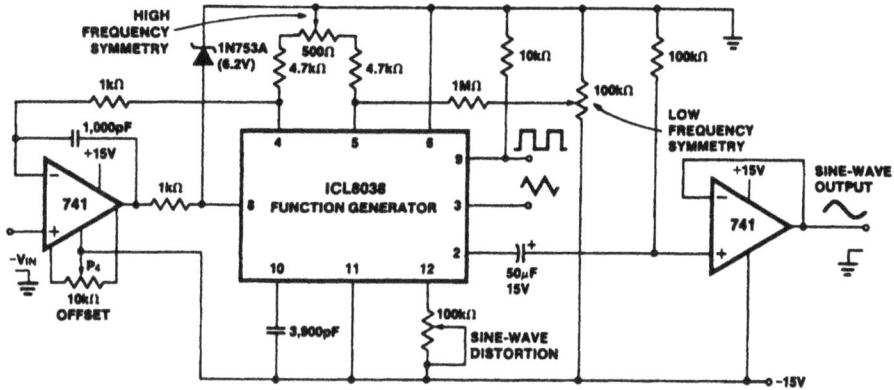

INTERSIL

Fig. 8-1

The linearity of input sweep voltage versus output frequency is significantly improved by using an op amp.

VOLTAGE-CONTROLLED OSCILLATOR (10 Hz To 10 kHz)

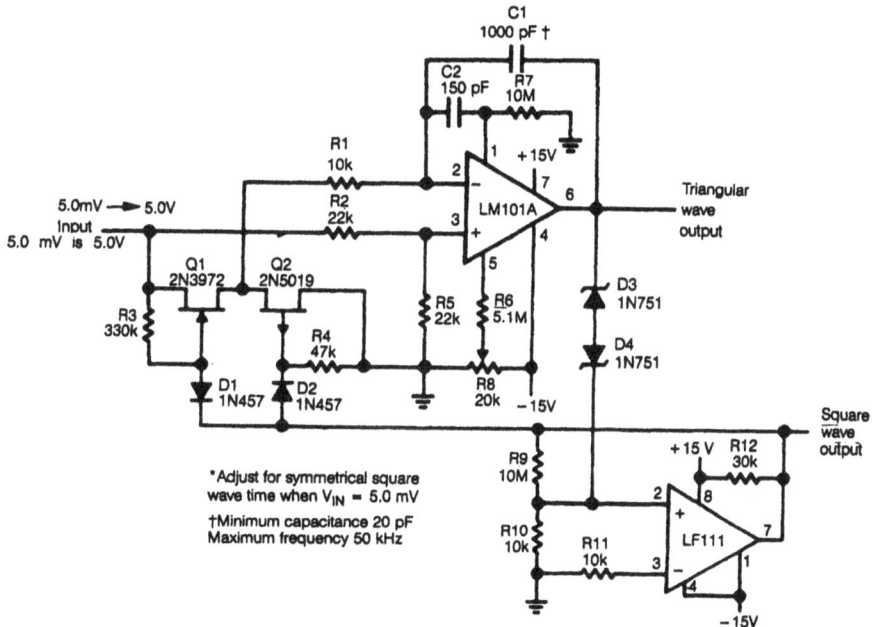

*Adjust for symmetrical square wave time when V_{IN} = 5.0 mV

†Minimum capacitance 20 pF
Maximum frequency 50 kHz

NATIONAL SEMICONDUCTOR

Fig. 8-2

VCO

Voltage-Controlled Oscillator Frequency vs Voltage

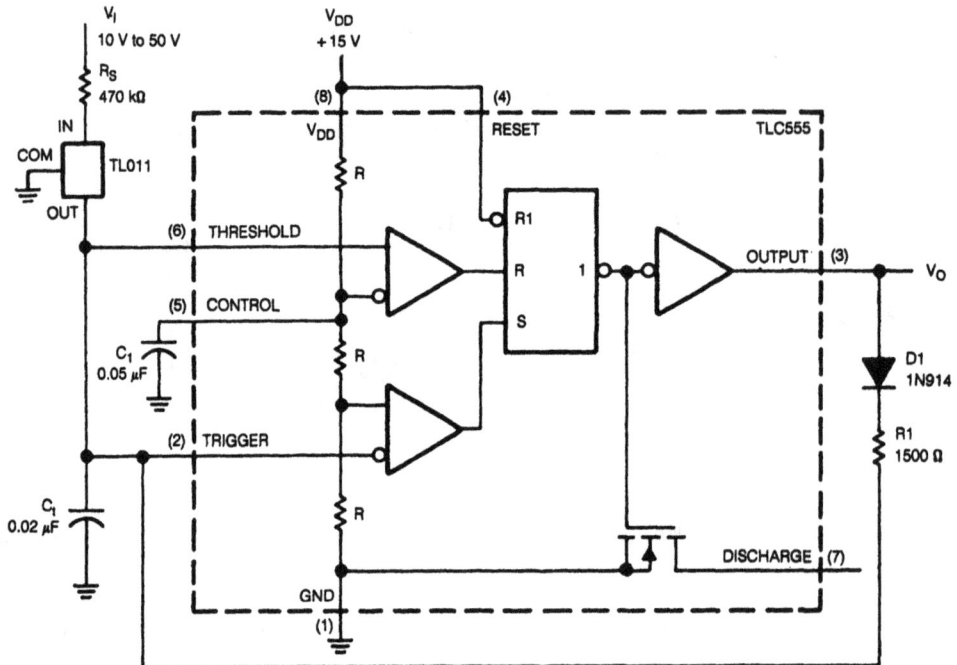

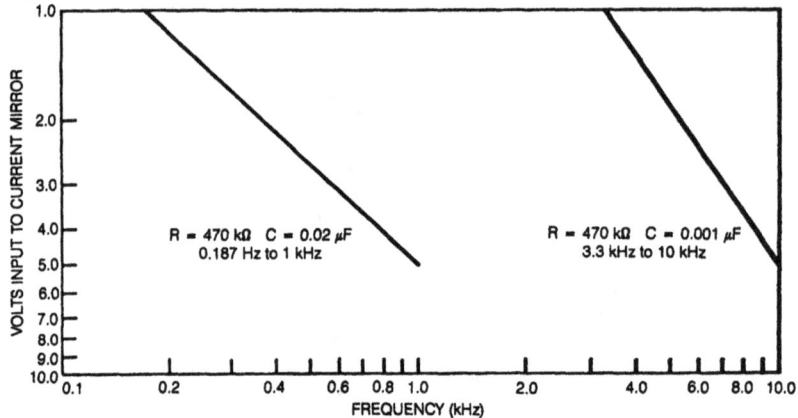

Fig. 8-3

At startup, the voltage in the trigger input at pin 2 is less than the trigger level voltage, $1/3\ V_{DD}$, causing the timer to be triggered via pin 2. The output of the timer at pin 3 becomes high, allowing capacitor C_t to charge very rapidly through diode D1 and resistor R1.

When capacitor C_t charges to the upper threshold voltage $2/3\ V_{DD}$, the flip-flop is reset, the output at pin 3 decreases, and capacitor C_t discharges through the current mirror, TLO11. When the voltage at pin 2 reaches $1/3\ V_{DD}$, the lower threshold or trigger level, the timer triggers again and the cycle is repeated.

SIMPLE VOLTAGE-CONTROLLED OSCILLATOR

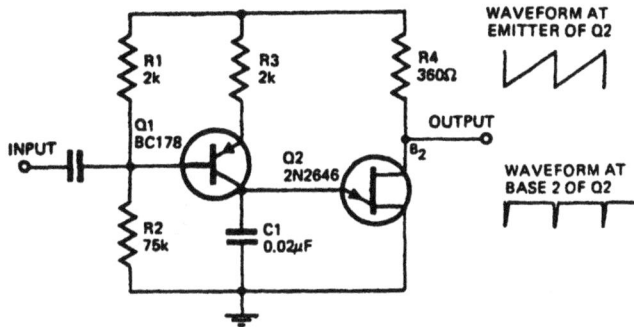

ELECTRONICS TODAY INTERNATIONAL

Fig. 8-4

With the component values shown, the oscillator has a frequency of 8 kHz. When an input signal is applied to the base of Q1 the current flowing through Q1 is varied, thus varying the time required to charge C1. As a result of the phase inversion in Q1, the direction of output frequency change is 180 degrees out of phase with the input signal. The output can be used to trigger a bistable flip-flop.

THREE DECADES VCO

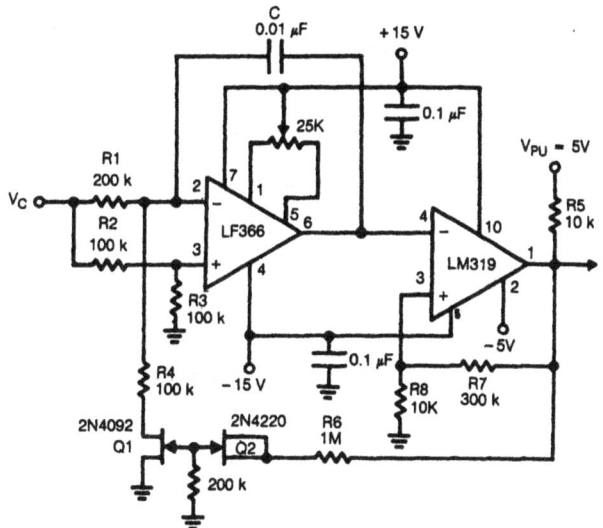

$$f = \frac{V_C(R8+R7)}{(8\,V_{PU}R8R1)C}, \quad 0 < V_C < 30V, 10\,Hz < f < 10\,kHz$$

R1, R4 matched. Linearity 0.1% over 2 decades.

NATIONAL SEMICONDUCTOR

Fig. 8-5

167

PRECISION VOLTAGE-CONTROLLED OSCILLATOR

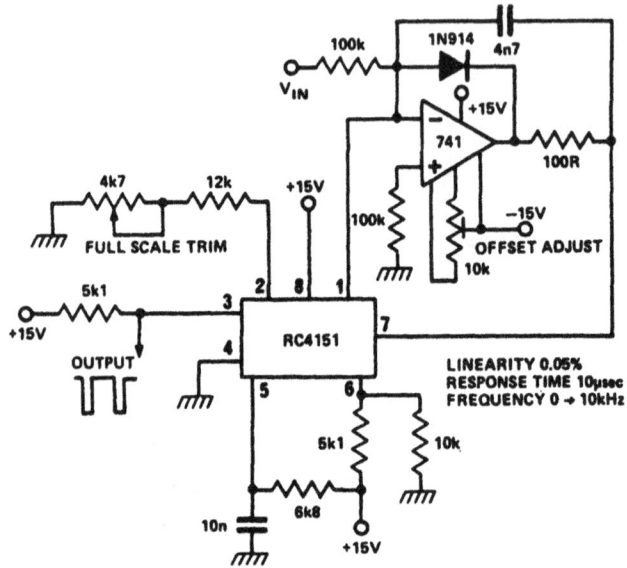

Fig. 8-6

RC 4151 precision voltage-to-frequency converter generates a pulse train output linearly proportional to the input voltage.

VOLTAGE-CONTROLLED OSCILLATOR

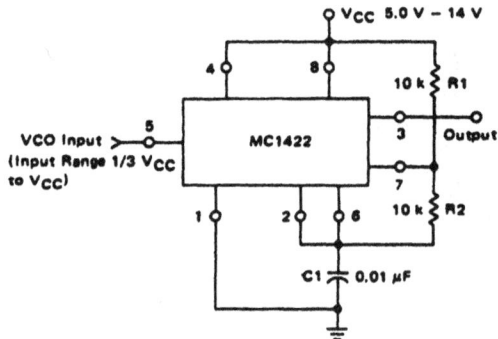

The VCO circuit, which has a nonlinear transfer characteristic, will operate satisfactorily up to 200 kHz. The VCO input range is effective from $1/3$ V_{CC} to V_{CC} -2 V, with the highest control voltage producing the lowest output frequency.

MOTOROLA

Fig. 8-7

WAVEFORM GENERATOR/STABLE VCO

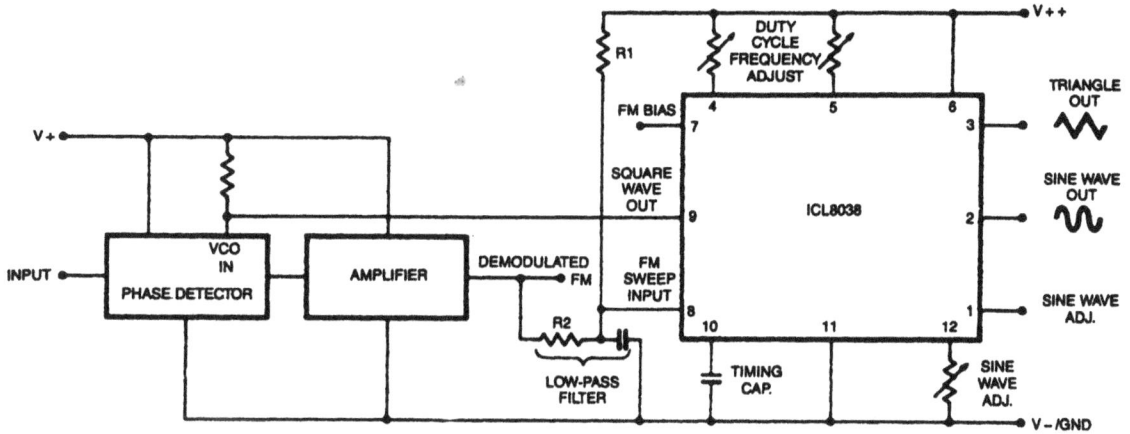

INTERSIL

Fig. 8-8

In this circuit, a waveform generator is used as a stable VCO in phase-locked loop (PLL).

VARIABLE-CAPACITANCE DIODE-SPARKED VCO

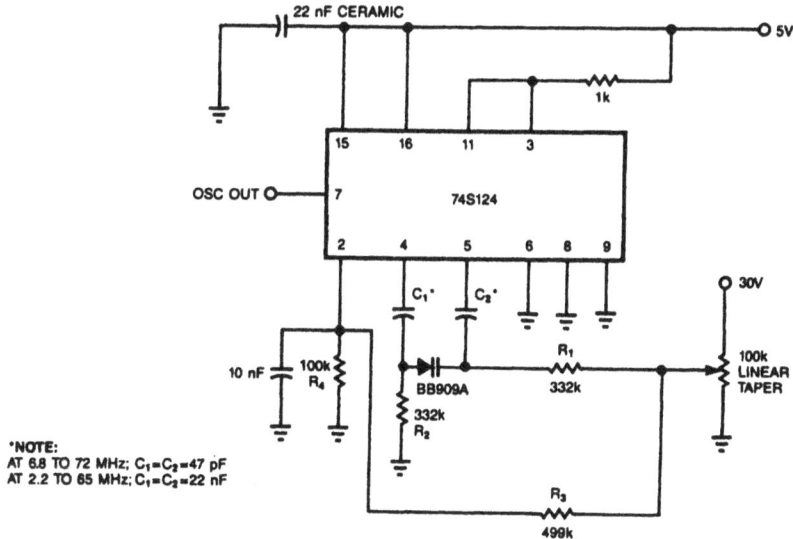

*NOTE:
AT 6.8 TO 72 MHz; $C_1=C_2=47$ pF
AT 2.2 TO 65 MHz; $C_1=C_2=22$ nF

EDN

Fig. 8-9

You can transform a 74S124 multivibrator into a wideband VCO by replacing its conventional fixed capacitor with a variable-capacitance diode. The only disadvantage of this scheme is the 30-V biasing voltage that the diode requires. Capacitors C1 and C2 couple the Philips BB909A variable-capacitance diode to the 74S124. R1 and R2 are large enough to isolate ground and control voltages from the timing capacitors. Resistors R3 and R4 form a voltage divider for the 74S124's control input.

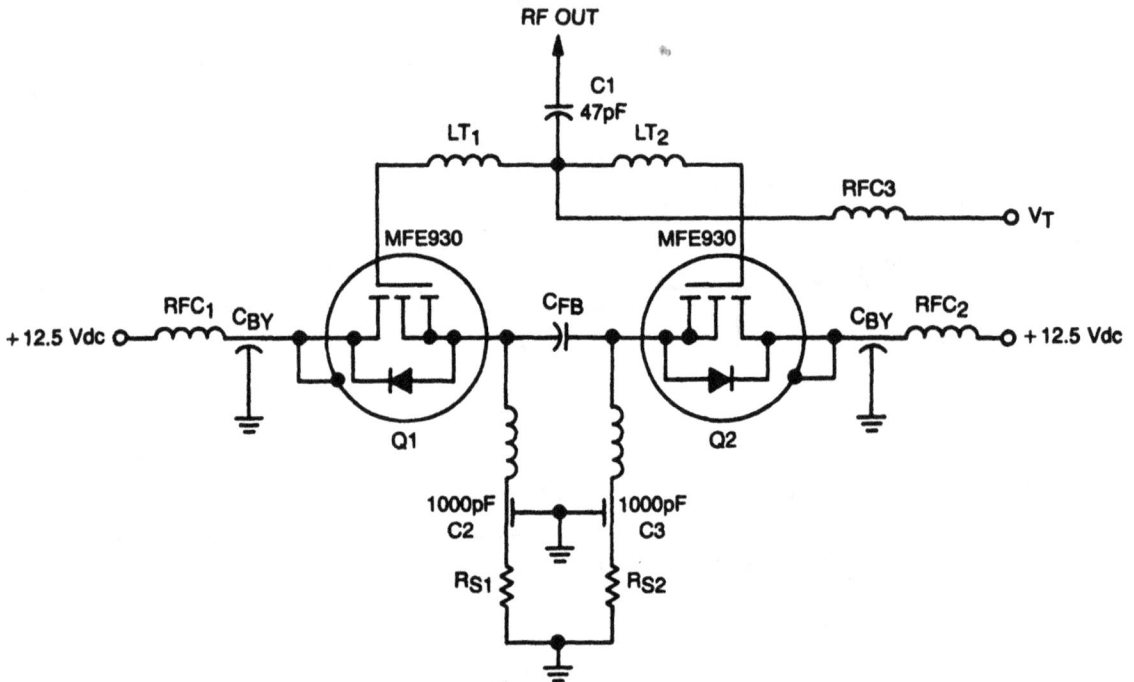

BALANCED TMOS VCO

Copyright of Motorola, Inc. Used by permission.

Fig. 8-10

This TMOS VCO operates in push-pull to produce 4 W at 70 MHz. It consists of two MFE930 TMOS devices in a balanced VCO that generally provide better linearity than the single-ended types. Varactors are not used because the design takes advantage of the large change in Miller capacitance, C_{RSS}, that is available in TMOS gate structures.

In the balanced VCO, the fundamental (f_o) and/or twice the fundamental ($2f_o$) can be coupled from the circuit at separate nodes. This makes the balanced oscillator very useful in phase-locked loops. The fundamental:

$$f_o = \frac{1}{2}(L_F C_{RSS})^{-1/2}$$

where:

$$L_F = 0.68 \ \mu H$$

TWO-DECADE HIGH-FREQUENCY VCO

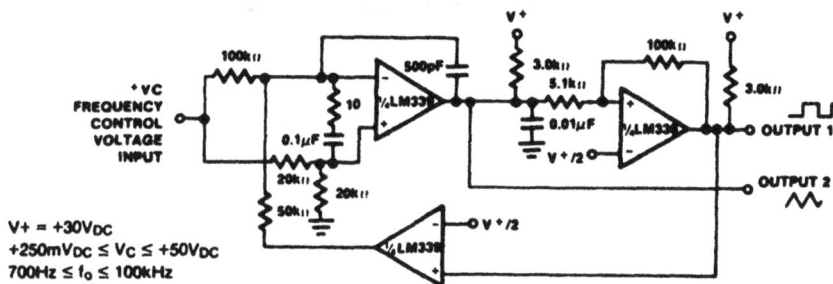

V+ = +30V_DC
+250mV_DC ≤ V_C ≤ +50V_DC
700Hz ≤ f_o ≤ 100kHz

SIGNETICS

Fig. 8-11

VOLTAGE-CONTROLLED OSCILLATOR

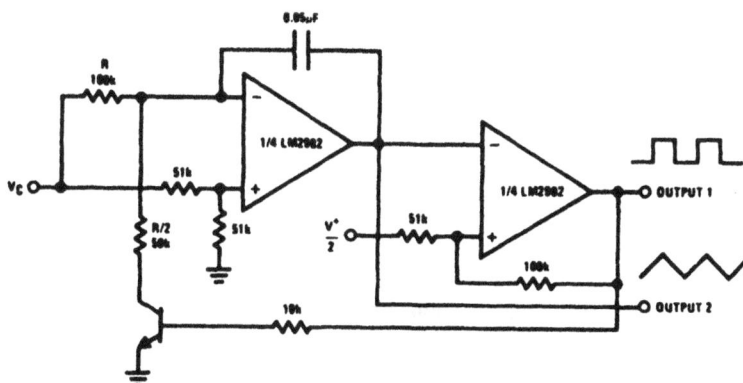

NATIONAL SEMICONDUCTOR

Fig. 8-12

VOLTAGE-CONTROLLED OSCILLATOR

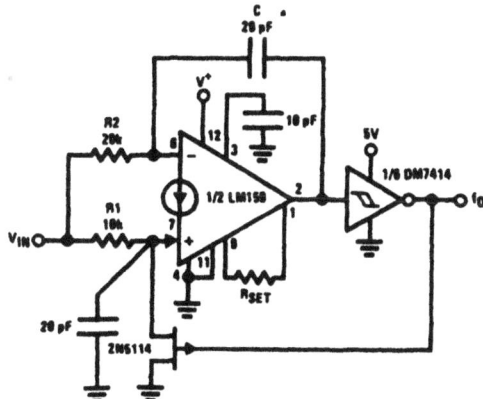

$$f_o = \frac{V_{IN} - \phi}{4C\Delta V\, R1}$$

where: R2 = 2R1

ϕ = amplifier input voltage = 0.6V

ΔV = DM7414 hysteresis, typ 1V

- 5 MHz operation
- T^{2}L ouput

NATIONAL SEMICONDUCTOR

Fig. 8-13

171

LOGARITHMIC SWEEP VCO

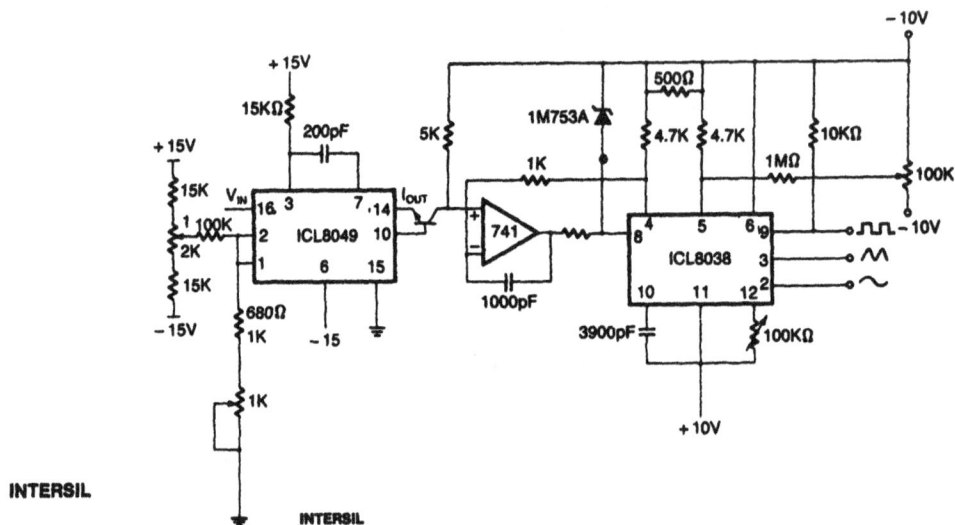

Fig. 8-14

This circuit uses the output of the ICL8049 to control the frequency of the ICL3038 waveform generator; the 741 op amp is used to linearize the voltage-frequency response. The input voltage to the 8049 can be, for example, from the horizontal sweep signal of an oscilloscope; the output of the 8038 will then sweep logarithmically across the audio range. By feeding this to the equipment being measured and detecting the output, a standard frequency response can be obtained.

SIMPLE VCO

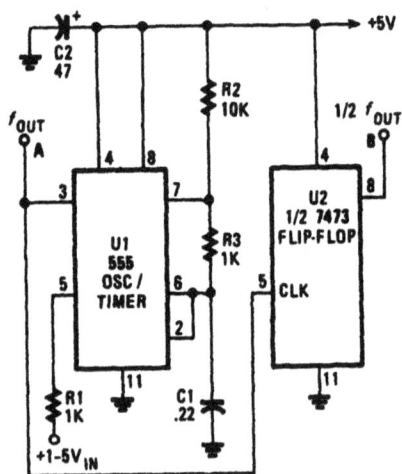

Fig. 8-15

The output frequency of the VCO, U1, varies inversely with the input voltage. With a 1-V input, the oscillator output frequency is about 1500 Hz; with a 5-V input, the output frequency drops to around 300 Hz. The output frequency range of U1 can be altered by varying the values of C1, R2, and R3. Increasing the value of any of those three components will lower the oscillator frequency, and decreasing any of those values will raise the frequency. Output-waveform symmetry suffers because the frequency varies from one extreme to the other. At the highest frequency, the waveform is almost equally divided. But when the frequency drops, the output of the circuit turns into a narrow pulse. If a symmetrical waveform is required, add the second IC, U2, half of a 7473P dual TTL J-K flip-flop, to the oscillator circuit.

PRECISION VOLTAGE-CONTROLLED OSCILLATOR

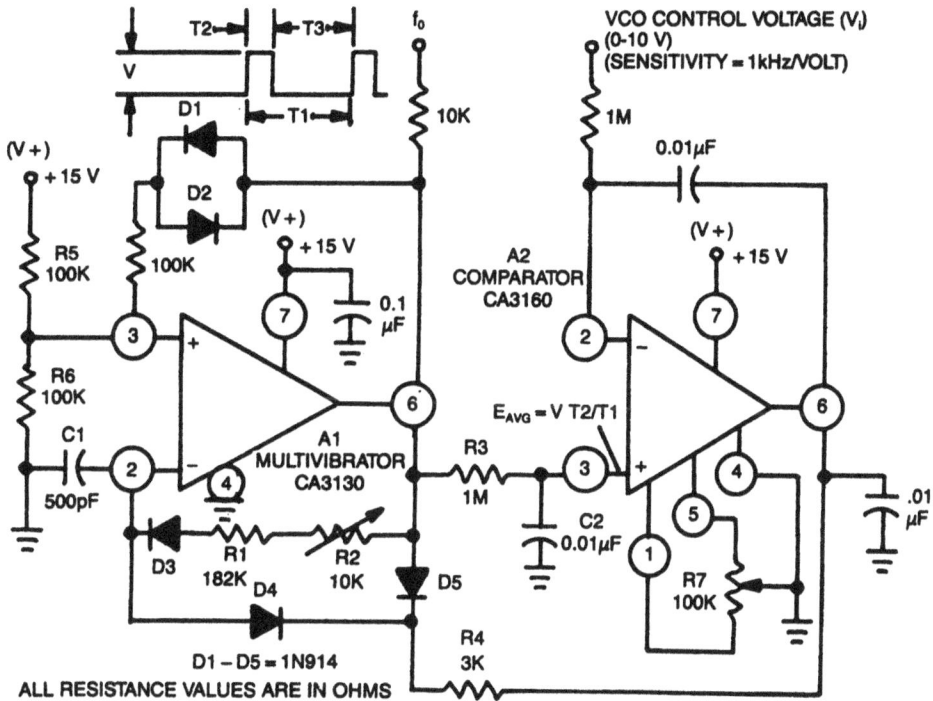

Fig. 8-16

This circuit uses a CA3130 BiMOS op amp as a multivibrator and CA3160 BiMOS op amp as a comparator. The oscillator has a sensitivity of 1 kHz/V, with a tracking error in the order of 0.02%, and a temperature coefficient of 0.01%/°C.

Sources

Chapter 1

Fig. 1-1. Reprinted with permission of National Semiconductor Corp., Transistor Databook, 1982, p. 11-31.

Fig. 1-2. Reprinted with permission of National Semiconductor Corp., Transistor Databook, 1982, p. 10-25.

Fig. 1-3. Courtesy of Fairchild Camera and Instrument Corp., Fairchild Semiconductor Application Note 300.

Fig. 1-4. Hands-On Electronics, Summer 1984, p. 43.

Fig. 1-5. How to Design/Build Remote Control Devices, TAB Book No. 1277, p. 230.

Fig. 1-6. Radio-Electronics, 7/83, p. 7.

Fig. 1-7. Ham Radio, 1/78, p. 78.

Fig. 1-8. Courtesy of Motorola Inc., Linear Integrated Circuits, 1979, p. 6-23.

Fig. 1-9. Electronics Today International, Summer 1982, p. 45.

Fig. 1-10. 73 Amateur Radio, p. 31

Fig. 1-11. Courtesy of Motorola Inc., Linear Integrated Circuits, 1979, p. 3-42.

Fig. 1-12. Reprinted with permission of National Semiconductor Corp., Linear Databook, 1982, p. 3-171.

Fig. 1-13. Reprinted with permission of National Semiconductor Corp., Linear Applications Handbook, 1982, p. AN29-9.

Fig. 1-14. Reprinted with permission of National Semiconductor Corp., Linear Applications Handbook, 1982, p. LB16-1.

Fig. 1-15. 73 Amateur Radio, 12/76, p. 97.

Fig. 1-16. 73 Amateur Radio, 7/77, p. 34.

Fig. 1-17. Popular Electronics, 10/89, p. 84.

Fig. 1-18. QST, 4/87, p. 48.

Fig. 1-19. Electronic Engineering, 5/85, p. 38.

Fig. 1-20. Electronic Design, 6/81, p. 250.

Fig. 1-21. Ham Radio, 1/87, p. 97.

Fig. 1-22. Electronic Engineering, 2/76, p. 17.

Fig. 1-23. Electronic Design, 2/73, p. 82.

Fig. 1-24. Radio-Electronics, 2/71, p. 37.

Fig. 1-25. Ham Radio, 6/82, p. 33.

Fig. 1-26. Linear Technology Corp., Linear Applications Handbook, 1987, p. AN5-7.

Fig. 1-27. Texas Instruments, Linear and Interface Applications, Vol. 1, 1985, p. 3-15 and 3-16.

Fig. 1-28. GE/RCA, BiMOS Operational Amplifiers Circuit Ideas, 1987, p. 7.

Fig. 1-29. Linear Technology Corp., Linear Databook Supplement, 1988, p. S2-34.

Fig. 1-30. Hands-On Electronics, 9/87, p. 97.

Fig. 1-31. Courtesy, William Sheets.

Fig. 1-32. Electronic Design, 10/65.

Fig. 1-33. Hands-On Electronics/Popular Electronics, 1/89, p. 26.

Fig. 1-34. Radio-Electronics, 8/86, p. 83.

Chapter 2

Fig. 2-1. Motorola, MECL System Design Handbook, 1983, p. 226.

Fig. 2-2. Reprinted from EDN, 2-78, © 1989 Cahners Publishing Co., a division of Reed Publishing USA.

Fig. 2-3. Third Book of Electronic Projects, TAB Book No. 1446, p. 22.

Fig. 2-4. Chrystal Oscillator Circuits, Robert. J. Matthys, Copyright © 1983, John Wiley & Sons, Inc., Reprinted by permission of John Wiley & Sons,Inc. RF Design 5-6/83, p. 69.

Fig. 2-5. Chrystal Oscillator Circuits, Robert. J. Matthys, Copyright © 1983, John Wiley & Sons, Inc., Reprinted by permission of John Wiley & Sons,Inc. RF Design 5-6/83, p. 64.

Fig. 2-6. The Complete Handbook of Amplifiers, Oscillators, and Multivibrators, TAB Book No. 1230, p. 330.

Fig. 2-7. Linear Technology Corp., Linear Databook, 1986, p. 2-104.

Fig. 2-8. QST, 12/85, p. 38.

Fig. 2-9. Ham Radio, 2/79, p. 41.

Fig. 2-10. The Complete Handbook of Amplifiers, Oscillators, and Multivibrators, TAB Book No. 1230, p. 329.

Fig. 2-11. The Complete Handbook of Amplifiers, Oscillators, and Multivibrators, TAB Book No. 1230, p. 336.

Fig. 2-12. 73 Amateur Radio, 8/78, p. 80.

Fig. 2-13. Electronics Today International, 8/73, p. 82.

Fig. 2-14. The Complete Handbook of Amplifiers, Oscillators, and Multivibrators, TAB Book No. 1230, p. 322.

Fig. 2-15. The Complete Handbook of Amplifiers, Oscillators, and Multivibrators, TAB Book No. 1230, p. 326.

Fig. 2-16. RF Design, 3/87, p. 31.

Fig. 2-17. 73 Amateur Radio, 1/89, p. 35.

Fig. 2-18. RF Design, 3/87, p. 31.

Fig. 2-19. Reprinted from EDN, 5/73, © 1989 Cahners Publishing Co., a division of Reed Publishing USA.

Fig. 2-20. NASA Tech Briefs, 8/89, p. 20.

Fig. 2-21. Linear Technology Corp., Linear Applications Handbook, 1987, p. AN3-14.

Fig. 2-22. Courtesy of Motorola Inc., Application Note AN-417B, p. 3.

Fig. 2-23. Courtesy of Motorola Inc., Application Note AN-417B, p. 5.

Fig. 2-24. Electronics Today International, 8/83, p. 57.

Fig. 2-25. Electronics Today International, 11/76, p. 44.

Fig. 2-26. Chrystal Oscillator Circuits, Robert J. Matthys, Copyright © 1983, John Wiley & Sons, Inc., Reprinted by permission of John Wiley & Sons Inc., RF Design, 5-6/83, p. 64.

Fig. 2-27. Chrystal Oscillator Circuits, Robert J. Matthys, Copyright © 1983, John Wiley & Sons, Inc., Reprinted by permission of John Wiley & Sons,Inc., RF Design, 5-6/83, p. 63.

Fig. 2-28. Chrystal Oscillator Circuits, Robert J. Matthys, Copyright © 1983, John Wiley & Sons, Inc., Reprinted by permission of John Wiley & Sons,Inc., RF Design, 5-6/83, p. 64.

Fig. 2-29. Ham Radio, 4/78, p. 50.

Fig. 2-30. Ham Radio, 2/79, p. 40.

Fig. 2-31. Ham Radio, 2/79, p. 42.

Fig. 2-32. The Complete Handbook of Amplifiers, Oscillators, and Multivibrators, TAB Book No. 1230, p. 330.

Fig. 2-33. The Complete Handbook of Amplifiers, Oscillators, and Multivibrators, TAB Book No. 1230, p. 331.

Fig. 2-34. Motorola, MECL System Design Handbook, 1983, p. 228.

Fig. 2-35. Motorola, MECL System Design Handbook, 1983, p. 227.

Fig. 2-36. Electronic Design, 11/69, p. 109.

Fig. 2-37. The Complete Handbook of Amplifiers, Oscillators, and Multivibrators, TAB Book No. 1230, p. 328.

Fig. 2-38. Signetics, 1987 Linear Data Manual Vol. 2: Industrial, 11/86, p. 5-269.

Fig. 2-39. Ham Radio, 2/79, p. 40.

Fig. 2-40. Ham Radio, 2/79, p. 42.

Fig. 2-41. Courtesy, William Sheets.

Fig. 2-42. Electronic Design, 10/75, p. 98.

Fig. 2-43. Reprinted from EDN, 8/7/86, © 1989 Cahners Publishing Co., a division of Reed Publishing USA.

Fig. 2-44. Reprinted from EDN, 6/83, © 1989 Cahners Publishing Co., a division of Reed Publishing USA.

Fig. 2-45. Radio-Electronics, 2/87, p. 96.

Fig. 2-46. Radio-Electronics, 2/86, p. 46.

Fig. 2-47. Reprinted from EDN, 6/83, © 1989 Cahners Publishing Co., a division of Reed Publishing USA.

Fig. 2-48. Maxim, Seminar Applications Book, 1988/89, p. 83.

Fig. 2-49. Courtesy, William Sheets.

Fig. 2-50. Linear Technology Corp., Linear Databook, 1986, p. 2-112.

Fig. 2-51. Reprinted from EDN, 10/1/87, © 1989 Cahners Publishing Co., a division of Reed Publishing USA.

Fig. 2-52. Ham Radio, 2/79, p. 43.

Fig. 2-53. Ham Radio, 2/79, p. 43.

Fig. 2-54. Ham Radio, 2/79, p. 43.

Fig. 2-55. Ham Radio, 2/79, p. 43.

Fig. 2-56. Intersil.

Fig. 2-57. The Complete Handbook of Amplifiers, Oscillators, and Multivibrators, TAB Book No. 1230, p. 324.

Fig. 2-58. Chrystal Oscillator Circuits, Robert J. Matthys, Copyright © 1983, John Wiley & Sons, Inc., Reprinted by permission of John Wiley & Sons, Inc., RF Design, 5-6/83, p. 63.

Fig. 2-59. Chrystal Oscillator Circuits, Robert J. Matthys, Copyright © 1983, John Wiley & Sons, Inc., Reprinted by permission of John Wiley & Sons, Inc., RF Design, 5-6/83, p. 63.

Fig. 2-60. Ham Radio, 2/79, p. 40.

Fig. 2-61. Chrystal Oscillator Circuits, Robert J. Matthys, Copyright © 1983, John Wiley & Sons, Inc., Reprinted by permission of John Wiley & Sons, Inc., RF Design, 5-6/83, p. 66.

Fig. 2-62. Ham Radio, 4/78, p. 50.

Fig. 2-63. Ham Radio, 2/79, p. 40.

Fig. 2-64. 73 Amateur Radio.

Fig. 2-65. Ham Radio, 4/78, p. 51.

Fig. 2-66. Modern Electronics, 6/78, p. 57.

Fig. 2-67. Ham Radio, 2/79, p. 39.

Fig. 2-68. Ham Radio, 3/82, p. 66.

Fig. 2-69. Ham Radio, 6/85, p. 23.

Fig. 2-70. Siliconix, MOSpower Design Catalog, 1/83, p. 5-27.

Fig. 2-71. Chrystal Oscillator Circuits, Robert J. Matthys, Copyright © 1983, John Wiley & Sons, Inc., Reprinted by permission of John Wiley & Sons, Inc., RF Design, 5-6/83, p. 63.

Fig. 2-72. Third Book of Electronic Projects, TAB Book No. 1446, p. 21.

Fig. 2-73. Reprinted with permission of National Semiconductor Corp., Linear Databook, 1982, p. 3-241.

Fig. 2-74. Teledyne Semiconductor Databook, p. 9.

Fig. 2-75. Reprinted with permission of National Semiconductor Corp., Application Note 32, p. 8.

Fig. 2-76. Reprinted with permission of National Semiconductor Corp., Transistor Databook, 1982, p. 7-26.

Fig. 2-77. Linear Technology Corp., Linear Applications Handbook 1987, p. AN20-12.

Fig. 2-78. QST, 2/78, p. 43.

Fig. 2-79. QST, 1/86, p. 40.

Fig. 2-80. Electronic Design, 11/74, p. 148.

Fig. 2-81. Ham Radio, 2/79, p. 40.

Fig. 2-82. RF Design, 3/87, p. 31.

Chapter 3

Fig. 3-1. Reprinted from EDN, 3/21/85, © 1989 Cahners Publishing Co., a division of Reed Publishing USA.

Fig. 3-2. Reprinted from EDN, 11/78, © 1989 Cahners Publishing Co., a division of Reed Publishing USA.

Fig. 3-3. Reprinted from EDN, 12/13/84, © 1989 Cahners Publishing Co., a division of Reed Publishing USA.

Fig. 3-4. Popular Electronics/Hands-On Electronics, 4/89, p. 23.

Fig. 3-5. GE/RCA, BiMOS Operational Amplifiers Circuit Ideas, 1987, p. 7.

Fig. 3-6. Electronic Engineering, 9/88, p. 34.

Fig. 3-7. Electronic Engineering, 9/88, p. 34.

Fig. 3-8. Reprinted from EDN, 4/14/84, © 1989 Cahners Publishing Co., a division of Reed Publishing USA.

Fig. 3-9. Electronic Engineering, 4/88, p. 33.

Fig. 3-10. Reprinted with permission from Electronic Design. Copyright 1989, Penton Publishing.

Fig. 3-11. Signetics, 1987 Linear Data Manual Vol. 1: Communications, 2/87, p.4-312.

Fig. 3-12. Texas Instruments, Linear and Interface Circuits Applications,1985, Vol. 1, p. 7-18.

Fig. 3-13. Intersil, 1978 Databook.

Fig. 3-14. Harris, Analog Product Data Book, 1988, p. 10-109.

Fig. 3-15. Siliconix Integrated Circuits Data Book, 1981, p. 8-51.

Fig. 3-16. Hands-On Electronics/Popular Electronics, 1/89, p. 84.

Fig. 3-17. GE/RCA BiMOS Operational Amplifiers Circuit Ideas, 1987, p. 9.

Fig. 3-18. National Semiconductor Corp., Linear Applications Databook, p.1118.

Fig. 3-19. Raytheon, Linear and Integrated Circuits, 1984, p. 12-8.

Fig. 3-20. Reprinted from EDN, 6/78, © 1989 Cahners Publishing Co., a division of Reed Publishing USA.

Fig. 3-21. Maxim, Seminar Applications Book, 1988/89, p. 45.

Fig. 3-22. Raytheon, Linear and Integrated Circuits, 1984, p. 12-7.

Fig. 3-23. Harris, Analog Product Data Book, 1988, p.

10-15.

Fig. 3-24. Intersil, Component Data Catalog, 1987, p. 7-104.

Fig. 3-25. Popular Electronics, Fact Card No. 98.

Fig. 3-26. Hands-On Electronics, Fact Card No. 86.

Fig. 3-27. GE/RCA, BiMOS Operational Amplifiers Circuit Ideas, 1987, p. 8.

Fig. 3-28. Hands-On Electronics, Fact Card No. 886.

Fig. 3-29. Harris, Analog Product Data Book, 1988, p. 10-168.

Fig. 3-30. GE/RCA, BiMOS Operational Amplifiers Circuit Ideas, 1987, p. 8.

Fig. 3-31. Raytheon, Linear and Integrated Circuits, 1989, p. 4-188.

Fig. 3-32. Reprinted from EDN, 6/78, © 1989 Cahners Publishing Co., a division of Reed Publishing USA.

Fig. 3-33. GE/RCA, BiMOS Operational Amplifiers Circuit Ideas, 1987, p. 8.

Fig. 3-34. Hands-On Electronics, Fact Card No. 89.

Fig. 3-35. Signetics, 1987 Linear Data Manual Vol. 1: Communications, 2/87, p. 4-311.

Fig. 3-36. Electronic Engineering, 9/84, p. 37.

Fig. 3-37. Electronic Design, 6/79, p. 122.

Fig. 3-38. NASA Tech Briefs, 6/87, p. 26.

Fig. 3-39. Electronics Today International, 6/80, p. 68.

Fig. 3-40. Electronic Engineering, 9/87, p. 27.

Fig. 3-41. Texas Instruments, Linear and Interface Circuits Applications, Vol.1, 1985, p. 7-25.

Fig. 3-42. Radio-Electronics, 5/70, p. 33.

Fig. 3-43. Linear Technology Corp, Linear Data Book, 1986, p. 5-78.

Fig. 3-44. Linear Technology Corp, Linear Data Book, 1986, p. 8-40.

Fig. 3-45. Electronic Engineering, 2/79, p. 23.

Fig. 3-46. Electronic Engineering, 7/86, p. 30.

Fig. 3-47. Motorola, Application Note AN-294, p. 6.

Fig. 3-48. Texas Instruments, Linear and Interface Circuits Applications, Vol.1, 1985, p. 7-16.

Fig. 3-49. Signetics, Analog Data Manual, 1982, p. 3-39.

Fig. 3-50. Motorola, Linear Integrated Circuits, p. 3-139.

Fig. 3-51. Hands-On Electronics, Winter 1985, p. 60.

Fig. 3-52. Reprinted from EDN, 1/77, © 1989 Cahners Publishing Co., a division of Reed Publishing USA.

Fig. 3-53. Intersil Component Data catalog, 1987, p. 7-44.

Fig. 3-54. Hands-On Electronics, Fact Card No. 89.

Fig. 3-55. GE/RCA, BiMOS Operational Amplifier Circuit Ideas, 1987, p. 10.

Fig. 3-56. Texas Instruments, Linear and Interface Circuits Applications, Vol.1, 1985, p. 3-20.

Fig. 3-57. National Semiconductor, Linear Brief 23.

Fig. 3-58. Texas Instruments, Linear and Interface Circuits Applications, Vol.1, 1985, p. 7-17.

Fig. 3-59. Signetics, Analog Data Manual, 1982, p. 8-10.

Fig. 3-60. Electronic Design, 11/29/84, p. 281.

Fig. 3-61. Electronics Today International, 12/78, p. 16.

Chapter 4

Fig. 4-1. Unitrode Corp., Databook 1986, p. 51.

Fig. 4-2. Electronic Engineering, 5/77, p. 27.

Fig. 4-3. National Semiconductor Corp., Transistor Databook 1982, p. 7-19.

Fig. 4-4. Electronics Today International, 12/78, p. 15.

Fig. 4-5. GE, Semiconductor Data Handbook, Third Edition, p. 513.

Fig. 4-6. GE/RCA, BiMOS Operational Amplifiers Circuit Ideas, 1987, p. 8.

Fig. 4-7. Linear Technology Corp., Linear Databook, 1986, p. 2-113.

Fig. 4-8. Reprinted from Electronics, 11/83, © 1983, McGraw-Hill, Inc., All rights reserved.

Fig. 4-9. Electronics Today International, 7/72, p. 84.

Fig. 4-10. GE, Application Note 90.16, p. 27.

Fig. 4-11. Elektor Electronics, 7-8/87 Supplement, p. 36.

Fig. 4-12. Harris, Analog Product Data Book, 1988, p. 10-15.

Fig. 4-13. Elektor Electronics, 7-8/87 Supplement, p. 63.

Fig. 4-14. Reprinted from EDN, 7/74, © 1989 Cahners Publishing Co., a division of Reed Publishing USA.

Fig. 4-15. Harris, Analog Product Data Book, 1988, p. 10-97.

Fig. 4-16. Linear Technology Corp., Linear Applications Handbook, 1987, p. AN20-11

Fig. 4-17. Fairchild Camera and Instrument, Linear Databook, 1982, p. 4-71.

Fig. 4-18. Reprinted from EDN, 8/76, © 1989 Cahners Publishing Co., a division of Reed Publishing USA.

Fig. 4-19. Reprinted from EDN, 5/11/89, © 1989 Cahners Publishing Co., a division of Reed Publishing USA.

Fig. 4-20. Reprinted from EDN, 4/82, © 1989 Cahners Publishing Co., a division of Reed Publishing USA.

Fig. 4-21. Hands-On Electronics, Fact Card No. 49.

Fig. 4-22. Reprinted from EDN, 2/2/89, © 1989 Cahners Publishing Co., a division of Reed Publishing USA.

Fig. 4-23. Reprinted from EDN, 1/78, © 1989 Cahners Publishing Co., a division of Reed Publishing USA.

Fig. 4-24. Harris, Analog Product Data Book, 1988, p. 10-108.

Chapter 5

Fig. 5-1. The Complete Handbook of Amplifiers, Oscillators, and Multivibrators, TAB Book No. 1230, p. 335.

Fig. 5-2. Courtesy of Fairchild Camera and Instrument Corp., Linear Databook, 1982, p. 9-28.

Fig. 5-3. Harris Semiconductor, Linear & Data Acquisition Products, 1977, p. 2-96.

Fig. 5-4. Electronics Today International, 7/78, p. 16.

Fig. 5-5. Courtesy of Motorola Inc., Linear Interface Integrated Circuits, p. 7-30.

Fig. 5-6. Reprinted with permission of National Semiconductor Corp., Data Conversion/Acquisition Databook, 1980, p. 13-50.

Fig. 5-7. Courtesy of Motorola Inc., Linear Interface Integrated Circuits, 1979, p. 7-9.

Fig. 5-8. Courtesy of Texas Instruments Inc., Linear Control Circuits Data Book, Second Edition, p. 145.

Fig. 5-9. Electronics Today International, 7/78. p. 16.

Fig. 5-10. Precision Monolithics Inc., 1981 Full Line Catalog, p. 8-31.

Fig. 5-11. Teledyne Semiconductor, Databook, p. 8.

Fig. 5-12. © Siliconix Inc., Application Note AN 154.

Fig. 5-13. Reprinted from Electronics, 2/77, p. 107. © 1985 McGraw-Hill Inc., All Rights Reserved.

Fig. 5-14. © Siliconix Inc., Analog Switch & IC Product Data Book, 1/82, p.6-19.

Fig. 5-15. Courtesy of Motorola Inc., Linear Interface Integrated Circuits, 1979, p. 6-136.

Fig. 5-16. Courtesy of Motorola Inc., Application Note, AN294.

Fig. 5-17. Courtesy of Fairchild Camera and Instrument Corp., Linear Databook,1982, p. 5-47.

Fig. 5-18. Signetics, 555 Timers, 1973, p. 22.

Fig. 5-19. Signetics, Analog Data Manual, 1983, p. 15-6.

Fig. 5-20. Precision Monolithics Inc., 1981 Full Line Catalog, p. 8-32.

Fig. 5-21. Courtesy of Fairchild Camera and Instrument Corp., Linear Databook, 1982, p. 5-46.

Fig. 5-22. Courtesy of Fairchild Camera and Instrument Corp., Linear Databook, 1982, p. 5-46.

Fig. 5-23. Reprinted with permission of National Semiconductor Corp., Linear Databook, 1982, p. 5-7.

Fig. 5-24. Reprinted from EDN, 4/30/87, © 1989 Cahners Publishing Co., a division of Reed Publishing USA.

Fig. 5-25. Reprinted from EDN, 5/28/87, © 1989 Cahners Publishing Co., a division of Reed Publishing USA.

Fig. 5-26. Exar, Telecommunications Databook, 1986, p. 9-24.

Fig. 5-27. Exar, Telecommunications Databook, 1986, p. 9-24.

Fig. 5-28. Courtesy of Texas Instruments Inc., Linear Control Data Book,Second Edition, p. 286.

Fig. 5-29. Courtesy of Texas Instruments Inc., Linear Control Data Book,Second Edition, p. 285.

Fig. 5-30. RCA Corp., Solid State Division, Digital Integrated Circuits Application Note, ICAN-6346, p. 5.

Fig. 5-31. Precision Monolithics In., 1981 Full Line Catalog, p. 16-154.

Chapter 6

Fig. 6-1. QST, 12/85, p. 39.

Fig. 6-2. Radio-Electronics, 7/83, p. 7.

Fig. 6-3. 73 Amateur Radio, 7/77, p. 35.

Fig. 6-4. 73 Amateur Radio, 11/85, p. 32.

Fig. 6-5. © Siliconix Inc., MOSPOWER Design Catalog, 1/83, p. 5-6.

Fig. 6-6. The Giant Book of Electronics Projects, TAB Book No. 1367.

Fig. 6-7. Reprinted with permission of National Semiconductor Corp., Linear Databook, 1982, p. 12-14.

Fig. 6-8. Radio-Electronics, 7/83, p. 7.

Fig. 6-9. Reprinted with permission of National Semiconductor Corp., Transistor Databook, 1982, p. 8-63.

Fig. 6-10. Reprinted with permission of National Semiconductor Corp., Transistor Databook, 1982, p. 11-32.

Fig. 6-11. Radio-Electronics, 5/70, p. 35.

Fig. 6-12. Reprinted from EDN, 1/79, © 1989 Cahners Publishing Co., a division of Reed Publishing USA.

Chapter 7

Fig. 7-1. Reprinted with permission of National Semiconductor Corp., Linear Databook, 1982, p. 3-289.

Fig. 7-2. Hands-On Electronics, Spring 1985, p. 35.

Fig. 7-3. Reprinted with permission of National Semiconductor Corp., Audio/Radio Handbook, 1980, p. 4-39.

Fig. 7-4. Reprinted with permission of National Semiconductor Corp., Linear Databook, 1982, p. 3-289.

Fig. 7-5. Electronics Today International, 11/76, p. 45.

Fig. 7-6. Electronics Today International, 2/75, p. 66.

Fig. 7-7. GE/RCA, BiMOS Operational Amplifiers Circuit Ideas, 1987, p. 28.

Fig. 7-8. Reprinted from EDN, 8/4/88, © 1989 Cahners Publishing Co., a division of Reed Publishing USA.

Fig. 7-9. Everyday Electronics, 5/88, p. 292.

Fig. 7-10. Hands-On Electronics, Winter 1985, p. 72.

Fig. 7-11. Reprinted with permission of National Semiconductor Corp., Audio/Radio Handbook, 1980, p. 4-29.

Fig. 7-12. Hands-On Electronics, 5-6/86, p. 86.

Fig. 7-13. Electronics Today International, 11/80, p. 43.

Fig. 7-14. Radio-Electronics, 2/75, p. 42.

Fig. 7-15. Electronics Today International, 1/77, p. 49.

Fig. 7-16. Courtesy, William Sheets.

Fig. 7-17. Radio-Electronics, 12/81, p. 53.

Fig. 7-18. Reprinted with permission from Electronic Design. © 1990, Penton Publishing.

Fig. 7-19. Electronics Today International, 1/77, p. 49.

Fig. 7-20. Popular Electronics, 8/74, p. 98.

Fig. 7-21. Popular Electronics, 12/74, p. 68.

Fig. 7-22. Electronics Today International, 1/77, p. 49.

Fig. 7-23. Electronics Today International, 1/77, p. 85.

Fig. 7-24. Electronics Today International, 6/75, p. 63.

Fig. 7-25. Courtesy William Sheets.

Fig. 7-26. Texas Instruments, Complex Sound Generator Bulletin No. DL-S 12612, p. 15.

Chapter 8

Fig. 8-1. Intersil Data Book, 5/83, p. 5-238.

Fig. 8-2. Reprinted with permission of National Semiconductor Corp., Linear Databook, 1982, p. 5-9.

Fig. 8-3. Texas Instruments, Linear Circuits Data Book, 1989, p. 2-73.

Fig. 8-4. Electronics Today International, 7/72, p. 84.

Fig. 8-5. Reprinted with permission of National Semiconductor Corp., Data Conversion/Acquisition Databook, 1980, p. 3-13.

Fig. 8-6. Electronics Today International, 12/78, p. 20.

Fig. 8-7. Courtesy of Motorola Inc., Linear Integrated Circuits, 1979, p. 6-17.

Fig. 8-8. Intersil Component Data Catalog, 1987, p. 6-27.

Fig. 8-9. Reprinted from EDN, © 1989 Cahners Publishing Co., a division of Reed Publishing USA.

Fig. 8-10. Motorola, Motorola TMOS Power FET Design Ideas, 1985, p. 47.

Fig. 8-11. Signetics, Analog Data Manual, 1982, p. 3-238.

Fig. 8-12. Reprinted with permission of National Semiconductor Corp., Linear Databook, 1982, p. 3-179.

Fig. 8-13. Reprinted with permission of National Semiconductor Corp., Linear Databook, 1982, p. 3-238.

Fig. 8-14. Intersil, Applications Handbook, 1988, p. 6-6.

Fig. 8-15. Popular Electronics, 10/89, p. 105.

Fig. 8-16. GE/RCA, BiMOS Operational Amplifier Circuit Ideas, 1987, p. 9.

Index

C